CLIMATE JUSTICE AND THE UNIVERSITY

CRITICAL UNIVERSITY STUDIES
Jeffrey J. Williams and Christopher Newfield, Series Editors

CLIMATE JUSTICE AND THE UNIVERSITY

Shaping a Hopeful Future for All

JENNIE C. STEPHENS

JOHNS HOPKINS UNIVERSITY PRESS | *Baltimore*

© 2024 Johns Hopkins University Press
All rights reserved. Published 2024
Printed in the United States of America on acid-free paper
9 8 7 6 5 4 3 2 1

Johns Hopkins University Press
2715 North Charles Street
Baltimore, Maryland 21218
www.press.jhu.edu

Library of Congress Cataloging-in-Publication Data is available.

A catalog record for this book is available from the British Library.

ISBN 978-1-4214-5005-6 (hardcover)
ISBN 978-1-4214-5006-3 (ebook)

Special discounts are available for bulk purchases of this book. For more information, please contact Special Sales at specialsales@jh.edu.

For Cecelia and Aden. And for all those converging in global solidarity for climate justice.

In memoriam:
Sinéad O'Connor and Professor Saleemul Huq

CONTENTS

List of Figures and Tables ix
Acknowledgments xi

INTRODUCTION. Another University Is Possible 1

1. Transformative Climate Justice in Universities 22
2. Injustices of Higher Education 58
3. Unlearning for Transformative Climate Justice 92
4. Exnovation Research and Knowledge Co-creation 125
5. Regenerative Financial Structures for Higher Education 152
6. Local Empowerment and Global Solidarity 187

CONCLUSION. Toward Climate Justice Universities 217

References 241
Index 275

FIGURES AND TABLES

FIGURES

1. Climate Justice Paradigm Shift 33
2. Knowledge, Wealth, and Power 41
3. White Male Leadership Still Dominates US Higher Education 68
4. Photos from Auroville, Tamil Nadu, India 101
5. The Financialized University 160
6. RISE Network Map 189

TABLES

1. Auroville Becoming 50 Program 99
2. Campus de la Transition's Pedagogical Areas 115
3. Values of the Ecoversities Alliance 195

ACKNOWLEDGMENTS

My two kids, who are now both adults, continue to be the people in this world that I learn from the most. I thank Cecelia Joy Stephens for her wisdom, compassion, joy, and enthusiasm. She has always been—and probably always will be—my most valuable and genuine critic. I thank Aden Sal Stephens for their intellectual clarity, generous spirit, and inner strength. The way they navigate the world with intentional care and curiosity is inspiring and makes me proud. Both Cecelia and Aden provided careful and insightful comments on sections of this book. I thank them both for their thoughtful feedback.

I thank both my parents, who are always so incredibly supportive in so many ways. My mother, Sarah Stephens, served as my editor throughout the writing process, and my father, Cathal Stephens, provided conceptual clarity and fact-checking as well as intellectual curiosity and inspiration. I also thank my siblings, Niall, Nora, and Kate, and their families for the wealth of care, love, and support that is always there. For inspiration and support, I thank Talaya Delaney and Arlene Delaney (may she rest in power). I also thank all of the young people in my life, especially my nieces and nephews (Mira, Cathal, Oisin, Rio, Caie, Eve, Nicholas, Theo, and Sarah), as well as so many inspiring students, who always provide me with grounding about what matters most in life.

So many colleagues and friends at different universities have contributed to my ideas in this book. I thank all my new colleagues at the National University of Ireland Maynooth and the ICARUS Climate Research Centre who have been supportive and so welcoming. From Northeastern University, I appreciate all of my colleagues within the School of Public Policy and Urban Affairs with whom I learned so much. I also thank many other Northeastern colleagues throughout the university, including those associated with the Women's, Gender, and Sexuality Studies program, the Global Resilience Institute, the Faculty Senate, the Climate Justice and Sustainability Hub, Northeastern London, the O'Bryant African American Institute, and Civil and Environmental Engineering.

I thank both Uta Poiger and Kathy Spiegelman, whose collaborative strategic approach to higher education leadership taught me so much. I also thank Mary Leffoletz for encouraging and supporting me as I explored academic leadership with a critical lens. I thank Ted Landsmark, who shared generously his wisdom and insights about many things, including university-community interactions. I also thank Rebecca Riccio for her collaborative and provocative approach to co-creating ethical principles for us all to engage with. Deep appreciation also to Regine Jean-Charles whose Black feminist leadership at Northeastern has expanded the space for imagining transformative change in so many impactful ways. Thanks also for the collaborative learning on institutional climate action with Leah Bamberg, Sheila Puffer, Diana Bozhilova, Dipa Desai, and Yutong Si. I also thank David Madigan for his frank and genuine responses to questions about multiple complex institutional challenges facing higher education.

Among the many, many students who I have engaged with in collaborative learning over the years, I am particularly appreciative of Alaina Kinol, who embraced our collaborative work on universities and led an amazing team of students to explore climate justice in higher education; her leadership resulted in the peer-reviewed article on climate justice in higher education published in *Climatic Change* that preceded this book. Thanks also goes to my impressive undergraduate research partners, Skye Lam, Saman de Silva, Yanyan Zeng, and Mellen Masea, whose contributions supported and informed the writing of this book.

I also thank Marla Perez-Lugo and Cecilio Ortiz-Garcia and the RISE Network, as well as Laura Kuhl and Ryan Ellis, who have been excellent collaborators on our Puerto Rico project. I appreciate the funding from the National Science Foundation for that research, a project that informed ideas in this book in multiple ways.

From the University of Vermont, I thank all my colleagues at the Gund Institute who introduced me to ecological economics before I fully appreciated how important it is. From Clark University, I appreciate my colleagues in the Department of International Development, Community, and Environment, which has now changed its name to the Department of Sustainability and Social Justice, including Halina Brown, Laurie Ross, and Ellen Foley. From Clark University, I also thank my mentors and friends in the Graduate School of Management, particularly Mary-Ellen Boyle and Ed Ottensmeyer from whom I learned so much.

Some of this book was written in Ireland while I was a visiting scholar at Trinity College Dublin, where I engaged with multiple colleagues who influenced my thinking. Specific thanks go to Anna Davies, Iris Moeller, Jane Stout, Susan Murphy, Clare Kelly, Linda Doyle, and Rory Rowan. Special appreciation to other Irish academic colleagues, including John Barry at Queen's University Belfast, who contributed in multiple ways to this book, including reading draft chapters. I also thank

Frances Fahy (University of Galway), Orla Kelly (University College Dublin), and Louise Fitzgerald (Dublin City University) for their impactful, fun, and productive collaborations.

This work has been influenced by interactions with many colleagues around the world. I thank Cécile Renouard at the Campus de la Transition in France and Manish Jain in India and other members of the Ecoversities Alliance. I also thank many collaborators and colleagues within the Climate Social Science Network (CSSN), especially those in the higher education working group and the politics of geoengineering working group. I also thank Maria Eugenia Hernandez for her lifelong friendship, inspiration, and collaboration on the role of universities in society, Derk Loorbach for his leadership in bridging the theory-to-action gap at Erasmus University in Rotterdam, and Elizabeth Wilson, whose conversations and friendship always inspire. And thanks to my PhD advisor, Janet Hering, who always encouraged expanding beyond conventional academic silos.

During the writing process, I was invited to speak at several universities and conferences in different contexts that helped me clarify my ideas. I am grateful to all those who engaged with me in this way and particularly grateful to those who invited me and organized the seminars.

I also thank all my creative colleagues who were at the Radcliffe Institute for the 2023–2024 academic year. I appreciate how, in different ways, everyone in that cohort was challenging the current state of knowledge and pushing boundaries. Special appreciation for Eddie Cole and Hazel Ruth Edwards, who I learned so much from as they were also exploring the power and constraints of higher education, and Fucia-Ann Hoover, who, like me, was navigating being a Climate Justice Fellow that year. Special thanks also to all those who were happy and proud to consider themselves "unreasonable."

This project has benefited from institutional support from the Harvard Radcliffe Institute for Advanced Study, Harvard's Salata Institute for Climate and Sustainability, the School of Public Policy and Urban Affairs at Northeastern University, and Northeastern London. Specific thanks also go to Frances Roberts-Gregory, Naomi Oreskes, Bill Clark, and Peter Frumhoff, who provided informal feedback and suggestions on this work.

I also want to acknowledge an extensive network of friends and family who I rely on in many different ways for love and support, including Driscoll friends, BHS friends, Minnesota women's leadership friends, and my many cousins, aunts, and uncles.

Special appreciation goes to my partner and collaborator, Martin Sokol, whose support, encouragement, love, and generosity cannot be described in words. Not only did he read and edit drafts of every chapter, but our many conversations have also inspired thoughtful connections that would never have emerged without him. His intellectual contributions to the evolution of my thinking about the role of universities in society are represented explicitly in figures 3 and 5. His 2003 PhD thesis provides the basis for the knowledge, wealth, and power framework (figure 3), and he contributed to improving the structure and content of figure 5.

Finally, I want to honor and appreciate the land and the nonhuman parts of earth's systems that provide us all with what we need to live, think, and contribute in our human societies. In particular, I honor the land, ocean, streams, wind, rain, plants, and birds of Meenacross in Donegal, where I was when I developed and wrote the most creative parts of this project.

INTRODUCTION

Another University Is Possible

During the first week of the first semester of my first faculty job, Hurricane Katrina devastated the US Gulf Coast. Video footage of the terrible human suffering in New Orleans exposed the racialized divides of climate vulnerabilities in the United States. It was August 2005, and I was an assistant professor of environmental science and policy at Clark University, a small private liberal arts research university west of Boston, Massachusetts, in the Northeast of the United States. Media coverage included disturbing images showing that most of those left behind struggling to survive after the hurricane were poor and Black. As students, faculty, and administrators reacted with dismay, compassion, and shock, I remember feeling an acute sense of inadequacy and disappointment that our university community was not able to do more to address the injustices of this destructive climate disruption. As hours turned into days, which turned into weeks and then months, the devastating scale of human suffering and displacement among those who had been systematically marginalized and underinvested in became clearer.

As the magnitude of the crisis unfolded, I was part of an innovative interdisciplinary academic department that focused on understanding issues of environment, community, and human displacement. We worked on conducting research and preparing students to engage with governments and organizations to reduce human and ecological vulnerabilities around the world. Despite this focus, my colleagues and I recognized that the financialized structures of US society that created the stark racial and economic disparities in human suffering during Hurricane Katrina were deeply entrenched. We also knew that the worsening climate crisis was exacerbating inequitable vulnerabilities and increasing the urgency to change those structures. Throughout that semester, our academic community continued to engage, discuss, and explore both the expanding disparate impacts and the societal responses to Hurricane Katrina.

I remember feeling both empowered and disempowered—hopeful and discouraged—at the same time. I was empowered and hopeful that the devastation would be a wake-up call—that the deep injustices exposed to the world after the hurricane would catalyze action. I thought the impetus to change the social and economic structures would minimize future suffering from worsening climate instability and devastating climate disruptions. But I also felt disempowered and discouraged, as it quickly became clear that the rebuilding and recovery efforts were disproportionately benefiting households and communities that were already better off before the hurricane. The response to the hurricane, like the responses to so many crises and climate disruptions, reinforced economic inequities and worsened racial disparities by providing more support to those who were already privileged and delivering less support to those most precarious and vulnerable (Tierney, Bevc, and Kuligowski 2006).

Throughout that first year as a young new professor, I became acutely aware of the minimal influence that our academic work had on the policies, practices, and systems that were creating, reinforcing, and expanding social and economic vulnerabilities as well as climate instability. Despite the idealism of our students and the commitment among the academic and professional staff within our university community to reduce human vulnerabilities exacerbated by the climate crisis, I recognized a disempowering acceptance of our limited role in making structural change toward climate justice. Like others, we could donate money, send supplies, or volunteer our time to travel to "help" those in crisis, but none of those compassionate responses would reduce the risks of the next devastating climate disruption. As academics, we could also continue to teach and conduct research to expose structural injustices and systemic deficiencies. But I found myself struggling to define what more we could do to create transformative change. How could we collectively leverage the power of higher education to shape a future with fewer Hurricane Katrinas? Could universities become critical infrastructure for advancing transformative climate justice by imagining and creating better futures for all?

For the past thirty years, my career within the higher education sector has been shaped by my continued efforts to respond to these questions. Throughout this time, as I have worked within several universities in the United States and spent time visiting universities in Europe, Asia, and the Middle East, I have been exploring the powerful potential of universities to respond to ecological destruction and social injustice. Through my teaching, research, and service—carried out in different ways with multiple initiatives, projects, and roles—I have collaboratively and collectively experimented with students and colleagues to expand the transformative societal impact of the work being done within higher education. While my sense of possibility for change has vacillated over these years, I have consistently maintained my fundamental belief that higher education is an underleveraged resource within society. Recognizing the powerful role that higher education plays in shaping the future, in this book I introduce a new concept: the idea of climate justice universities. Climate justice universities are higher education systems committed to addressing the planetary crisis by shaping more stable, healthy, and equitable futures for all. Imagining the possibilities of climate justice universities requires a paradigm shift in thinking about the role and influence of higher education in society. With climate chaos exacerbating injustices and instability of all kinds and worsening the other intersecting crises facing humanity, reimagining the transformative power of higher education is a hopeful and empowering initiative.

Acknowledging the powerful potential of universities, this book invites readers to reimagine how higher education systems could be reclaimed and restructured to better prepare society for a hopeful future. While most universities have made commitments to engage with issues of sustainability and climate change, many of these efforts are limited and constrained, and few include an orientation toward transformative social and economic change. Sustainability in higher education has tended to focus on incremental improvements in reducing waste and improving efficiency, and university climate action plans are often limited to strategies to reduce greenhouse gas emissions on campus. Universities are notoriously resistant to change because academic work tends to reinforce disciplinary boundaries and uphold institutional traditions. Engaging with transformative social change is often considered to be misaligned with conventional academic procedures and priorities. Courage is required, therefore, to let go of assumptions and reimagine higher education. The scale of internal resistance to change within universities is illustrated by this anonymous quote: "*Changing a university is like moving a graveyard—you don't get much help from the inside.*" Because of the internal pressure within higher education to maintain institutional norms, this book and its proposal for climate justice universities are, in some ways, radical acts of resistance. The idea of reclaiming universities as critical infrastructure to advance transformative social change to create better futures for all could even be viewed as revolutionary.

With communities around the world struggling to learn how to adapt to climate instability and growing vulnerabilities (Favretti 2023), most universities are not yet transforming their approach to adapt to better serve society. In many places, higher education institutions are shifting in the opposite direction, narrowing their opportunities and becoming more exclusive in terms of what communities they serve and what kind of research they conduct. During this destabilizing time, the mission of universities to serve the public and prepare students for the future seems to have been diluted and diminished while higher education institutions are being increasingly leveraged to further concentrate wealth and power. As large corporations and the ultrarich gain increasing influence within society, higher education systems are in many places increasingly catering to the preferences and priorities of wealthy and powerful interests.

Another university is possible however. Rather than continuing to reinforce an exclusive and individualistic approach that links academic and financial success, higher education could adopt a more inclusive and collective approach that prioritizes the common good and shifts human societies onto a path toward a more stable and healthy future for all. Given growing climate destabilization, reimagining the potential of higher education is both necessary and urgent. If transformative changes are not considered in the higher education sector, universities are at risk of continuing to legitimize the dehumanizing and destabilizing social and economic systems that are no longer serving humanity. Higher education institutions themselves may be increasingly considered unfit for purpose.

Collective Reimagining

This book is an invitation to collectively reimagine how higher education systems could be reclaimed and restructured for the public good. What if all universities were climate justice universities pursuing the collective mission of moving society onto a path toward a more equitable, just, ecologically healthy, climate-stable future? Here, I make the case that higher education has a huge, but largely untapped, potential for supporting and accelerating the transformative social, political, and economic changes required to maintain a society with human well-being and planetary health at its core. I suggest that this potential can be realized only by resisting the capture of higher education by powerful elites and corporate interests and disrupting the ways in which universities concentrate wealth and power by gatekeeping knowledge. By interrogating the insidious links among knowledge, wealth, and power, an invigorated focus on reclaiming higher education to advance the common good emerges. This book explores possibilities for restructuring universities as institutions committed to redistributing and sharing knowledge, wealth, and power rather than perpetuating policies and practices that concentrate and constrain knowledge, wealth, and power. In many contexts around the world, this requires a paradigm shift. For universities to become accessible, regenerative, and reparative—that is, institutions focused on locally engaged collaborations, global solidarity, and inclusive and creative learning experiences—a transformative vision is needed. This alternative vision of climate justice universities calls for collectively resisting and discarding the legacy and traditions of exclusivity, competition, exploitation, and self-aggrandizement that currently characterize many higher education systems throughout the world.

This book contributes to the expanding field of critical university studies, which reveals and critiques the societal harms associated with corporatization, precarious academic labor, ballooning student debt, and other extractive and exploitative practices within higher education (Boggs and Mitchell 2018; Williams 2021). This book encourages prioritizing collective action, civic engagement, and ecological health rather than maintaining a narrow focus on promoting individual and institutional success by striving to maximize the financial status of the university and its students, faculty, staff, administrators, and alumni.

This book reimagines universities and their potential to advance societal transformation by considering three domains of system change: resisting, reclaiming, and restructuring. I adapt these three categories from my research on energy democracy, which refers to the idea that the transformation away from fossil fuels toward a renewables-based future provides opportunities not just to redistribute electric power but also to redistribute economic and political power for a more just and equitable society. While energy democracy focuses on resisting fossil fuels, reclaiming energy decision-making, and restructuring energy systems for more equitable and renewable power (Burke and Stephens 2017), transformation in higher education requires *resisting* the financialization of higher education and the forces that prioritize private, individual benefits instead of collective benefits; *reclaiming* higher education to prioritize human well-being, ecological health, and the common good; and *restructuring* the goals of higher education toward shaping more equitable, climate-stable, sustainable, and healthy futures for all.

My primary intention in writing this book and proposing a paradigm shift toward climate justice universities is to stimulate new ideas and different conversations about the future of higher education. I hope to expand our imaginations about what is possible and inspire new ways of thinking about universities. I recognize this is a provocation, so I anticipate that some may find these ideas unrealistic or too radical. The power of imagination, however, allows us to reconsider our current assumptions and expand our view of what is possible (Benjamin 2024). One approach that I find helpful in trying to interpret diverse perspectives is to recognize that contrasting views are related to different assumptions about what aspects of society are fixed, rigid, and will never change and what parts of society are flexible, adaptable, and possible to change. As someone with an active imagination for change and a strong instinct to resist getting stuck, I believe deeply that transformative social change toward climate justice and collective well-being is not only possible but that it is likely. As the ecological conditions supporting humanity continue to deteriorate, expanding social movements around the world are rising up, demanding change, and forming new global coalitions. In solidarity with so many creative and inspiring advocates for social and economic justice, I offer these ideas about climate justice universities with hope and humility.

My goal for proposing this new concept of climate justice universities is to catalyze a paradigm shift in how we collectively consider the power of higher education institutions. I hope that this concept will resonate with a broad range of people within and outside formal education systems around the world. From idealistic, angry, and disillusioned youth to conservative senior university administrators, from activists and policymakers to teachers, students, professors, and researchers, I hope these ideas about climate justice and the university are inspiring and empowering. I recognize that some readers may argue that the mission of higher education should not be entangled in politically charged issues and, thus, universities should not be oriented toward social change, social justice, and transformation. I welcome this critique as it catalyzes important debate, discussion, and intentionality about the role of higher education in society and how universities can or should engage or disengage. Given that universities are already deeply entangled, both politically and financially, in shaping climate futures, higher education cannot pretend to be neutral when it comes to climate justice. I understand that some readers may feel threatened or unsettled by the ideas and assumptions in this book. I also anticipate that others will feel supported, connected, and inspired. Because transformative change requires disruption, tension and discomfort are inevitable when transformative ideas are explored.

A Justice-First Perspective

Since I began my academic career, my environmental interests at the interface of science, engineering, and policy have become increasingly focused on equity, solidarity, and justice. Building on my initial interest in environmental science and technology, I have come to realize that humanity's collective inability to effectively respond to the worsening climate crisis—a crisis that is exacerbating all other vulnerabilities—results not from a lack of science or technological advances but from the concentration of wealth and power that relies on sustained exploitation of people and the planet and constrains knowledge and our collective imaginations about social change. To perpetuate wealth accumulation, powerful elites and those profiting from fossil fuels and other extractive industries have invested for decades to limit knowledge production and constrain knowledge dissemination related to our collective societal responses to climate change. A narrow focus on profit and growth above all else has distorted societal priorities, reinforcing a capitalistic system that is devastating the health of people and the planet. Higher education institutions have been manipulated with investments to intentionally spread misinformation and to advance a simplistic, constrained individualism that undermines democratic deliberative processes and obstructs policies to move humanity toward a more equitable and climate-stable future. Higher education institutions are not only complicit in these exploitative processes, they play an increasingly central role in reinforcing authoritative, extractive capitalism and squashing alternative visions of the future.

From the first time I developed and taught a university-based course linking climate science and policy with social change and social justice in 2004 at Tufts University's "Experimental College" (outside of Boston in Somerville, Massachusetts) to my decades of collaborative research oriented toward accelerating transformation by understanding the barriers to energy system change and the phasing out of fossil fuels, I have experimented with novel approaches to teaching, learning, and research. My primary goal has been to contribute to social change toward a more just and equitable world. It was not until my four years in an administrative role at Northeastern University in Boston, where I served as the director of the School of Public Policy and Urban Affairs, that I realized the magnitude of restrictions within the higher education system that constrain how universities—and the people associated with them—can engage in collective structural and systemic change for the public good. During this time, I learned that in increasingly financialized capitalistic societies, where the concentration of wealth and power is widely accepted, applauded, and promoted, many higher education institutions are perpetuating and reinforcing the extractive and oppressive systems of debt and dehumanization that enable the growth of a rapidly expanding billionaire class. And it is this concentration of wealth and power—the rise of the billionaire class—that is thwarting transformation toward a more just, equitable, and climate-stable future (Kenner 2019). Although some billionaires donate money to universities and contribute to environmental causes, the lack of society-wide appreciation of the economic and political injustices of this concentration of wealth and power is among the biggest obstacles to change (Kelly 2023).

My career in higher education has opened my eyes to how universities play a major role in endorsing and legitimizing cultural complacency about growing political, economic, and environmental injustices. While universities around the world publicly commit to diversity, equity, and inclusion; advancing the public good; and addressing the grand challenges facing humanity, many are stuck in a system in which donors and politicians actively constrain what is taught, researched, and discussed on campus. Despite the ideal of open inquiry and dialogue, most universities are compelled to respond to the preferences and priorities of those who control their funding, even when it requires limiting academic inquiry in some areas while promoting other areas.

I am continually inspired by many dedicated academic colleagues and students who are committed to dismantling systems of injustice and oppression and steering humanity toward a more healthy, stable, and equitable future. Universities provide critically important spaces to challenge exploitative power structures and confront humanity's biggest challenges. But as the entangled polycrises of climate, war, mental health, and economic precarity rapidly expand, it is increasingly clear to many that higher education could be doing much more to halt the growing instability, destruction, and suffering. While calls for institutional neutrality reinforce current systems and structures, institutional actions that explicitly resist continued injustice and confront oppression are urgently needed for the deep societal transformations required to stabilize the climate and reverse growing economic precarity. Given that financial entanglements of universities have always influenced the academic priorities of specific institutions, there is no such thing as institutional neutrality in higher education. As American historian and social justice activist Howard Zinn declared in the title of his 2002 memoir, *"You can't be neutral on a moving train."* Transparency and institutional self-reflection are, therefore, essential to enable open academic inquiry and accountability during this time of accelerating climate injustices. Institutional actions and commitments are necessary not only to amplify the actions and commitments of individuals but also to resist and prevent universities from being captured and co-opted by wealthy and powerful interests. A new era of intersectional, cross-sectoral, international collective coalition work is emerging around the world to disrupt structural injustices and co-create a path toward a more healthy and regenerative future based on a solidarity, well-being economy (Matthaei and Slaats 2023). The opportunity ahead is for higher education institutions to engage in this collective work by reimagining and redefining their public good mission as universities striving for climate justice.

In my 2020 book, *Diversifying Power: Why We Need Antiracist, Feminist Leadership on Climate and Energy*, I argue that the scale of change that is needed for climate stability and climate justice is not possible until and unless power and voice are given to more diverse leadership representing different experiences and priorities. The world needs more leaders who recognize how the biggest challenges facing society are linked and that the best opportunities for change are when these challenges are addressed together. In that book, which was published in the early months of the COVID-19 pandemic, I honor many creative and inspiring leaders who are connecting climate and energy with economic justice and jobs, health equity, food justice and ecological agriculture, transportation justice, and housing for all. While writing *Diversifying Power*, at one point I wanted to add a chapter linking climate and energy leadership to education, but my excellent editor (Heather Boyer at Island Press) dissuaded me from doing this, suggesting that it was a topic that deserved more than a single chapter. So here, in this book, I build on the importance of diversifying power and focus on the transformative potential of higher education to advance a more just and stable future.

I recognize the limitations and biases of my perspective as a white woman with the privileges associated with being born in Ireland into a large supportive family that prioritizes education, equity, and compassion. Raised within a culture that proudly embraces its long legacy of resistance to colonial oppression, I am aware of how my background and life experiences have oriented me toward challenging the powerful forces that encourage complacency to injustices. This book embraces a justice-first perspective, which, to me, means resisting extractive and exploitative power structures that give advantage to some by exploiting the disadvantage of others. A justice-first perspective is an antiracist, feminist, decolonial perspective. A justice-first perspective acknowledges that in the collective struggle for justice and liberation, there is no place for complacent acceptance of the injustices of the status quo.

Linking Universities with Global Solidarity and Climate Justice

To explore the idea of climate justice universities, throughout this book I connect my own ideas and experiences reimagining the powerful potential of higher education with the larger transformative ideas of global solidarity and climate justice that are expanding around the world (Sultana 2022a; Matthaei and Slaats 2023). Global solidarity means sharing in struggle, embracing a collective worldview of humanity, and recognizing that all people need and deserve the same basic necessities in life. Solidarity is essential for community; it is that fundamental sense of belonging and being connected to others (Matthaei and Slaats 2023). Climate justice means focusing on transformative social, political, and economic changes to reduce inequities and disparities in health, wealth, and vulnerabilities (Newell et al. 2021). Climate justice also acknowledges the extremely uneven and inequitable impacts of climate change; climate action that does not center justice and equity encourages the concentration of wealth and power and worsens climate vulnerabilities among those who are already suffering the most (Stephens 2022). Climate justice acknowledges the colonial legacy of the uneven distribution of suffering among those who have contributed the least (Sultana 2022b).

Communities around the world are facing both worsening economic precarity and increased suffering of all kinds from climate instability. In response to a growing sense of dispossession, disconnection, and disruption, new emerging coalitions are expanding a climate justice movement based on global solidarity and feminist principles (Sultana 2022b). Rather than accepting the dominant, cis-heteropatriarchal, capitalistic hierarchies that are reinforcing economic injustice and climate chaos by privileging those who are already privileged and disadvantaging those who are already disadvantaged, this climate justice movement is connecting struggles for liberation, justice, and peace with resistance to capitalistic extraction of fossil fuels and exploitation of people and communities (Sultana 2022a). As part of the growing climate justice movement, transformative proposals for new ways of structuring societies are gaining traction, including alternative economic systems like the solidarity economy (Matthaei and Slaats 2023), the caring economy (Lorek, Power, and Parker 2023), the post-growth economy (Jackson 2022), the well-being economy (Hayden and Dasilva 2022), the democratic economy (Kelly 2023), and other proposals for regenerative rather than extractive societies (United Frontline Table 2022).

Since Hurricane Katrina in 2005, climate suffering has worsened, and it is increasingly clear that society's efforts to reduce the risks of climate change have so far been inadequate and insufficient. Humanity is moving in the wrong direction, increasing rather than decreasing greenhouse gas emissions and expanding rather than reducing economic inequities, precarity, and climate vulnerabilities. Despite more than thirty years of international climate negotiations organized by the United Nations Framework Convention on Climate Change, fossil fuel companies continue to expand oil and gas exploration and extraction, making record profits with no plans (and no regulatory incentives) to stop. Climate-disruptive events, including droughts, floods, fires, heat waves, and storms, are becoming more frequent and intense, and marginalized households, communities, and regions throughout the world are suffering the most.

During the past thirty years, higher education systems throughout the world have been expanding, and access to higher education in many regions of the world has increased considerably (UNESCO 2020). Despite these advances, most university systems are not yet engaging with and preparing for the rapidly destabilizing future ahead. By linking their educational and research missions with global solidarity and climate justice, higher education institutions could expand their impact and relevance by connecting more directly with a diversity of community needs.

Starting Points

Reimagining higher education within the paradigm of climate justice universities may be difficult because the realities of many contemporary higher education institutions are so far from these ideals. But the first step in implementing structural changes to respond to the intersecting crises facing humanity is creating a new vision. Most of us are ill-equipped to imagine transformative change, and many are unwilling to even try. But the hardest part of transformative change is not identifying the deficits of the current systems, nor is it developing potential solutions to specific problems. The biggest challenge to systemic transformation is the constraint imposed by our current understanding of our starting point. Each individual's perception of the future, their understanding of what is possible and what is impossible, is based on individual and collective perceptions about where we are now. Our anticipation of what is likely or unlikely to happen in the future is based on where we think we are currently. And where we each think we are right now emerges from a synthesis of our collective life experiences; that is, we are all constrained and shaped by our own realities and the realities of our communities. Rather than continuing to be stuck limiting our expectations of the future by accepting our current starting points, I suggest it is time to take a leap by assuming a different starting point to expand our collective imaginations. With the growing uncertainties about the future in this era of more frequent intense climate disruptions, accepting our default starting points will surely continue to constrain and disempower us. Instead of feeling overwhelmed or discouraged by the magnitude of change that is needed when we view the future from our current starting point, this book invites us all to imagine starting from a new and different place.

The constraints of our current starting points evoke an old Irish joke about a tourist navigating the Irish countryside looking for a specific place. When the tourist stops to ask for directions, a local farmer on the side of the road says, "Well, sir, if I were you, to get there I wouldn't start from here." The point being that if you know where you want to go, it may be better to start from a place different from where you are right now. Different starting points offer different likelihoods of reaching the destination; choosing a different starting point allows us to see alternative ways of moving toward our destination.

This book is an invitation to envision climate justice universities as a new destination, a place we have not previously traveled to. This book acknowledges that although the path to get to this reimagined different kind of higher education system is not clear, creating the vision for where we want to go is the first step. To enable clarity as we imagine this preferred destination, we can challenge ourselves to start from a place different from where we typically start. Rather than accepting the constraints of where we find ourselves in our current situations, what if we each let go of our current positionality and joined in a collective reimagining of a different kind of university. Shifting starting points will be easier for some than for others; readers will all be at different levels of embeddedness within and outside contemporary higher education systems. I suggest that changing the starting point is liberating and empowering for everyone because it unleashes imagination, reveals obscured realities, and sparks new ways of thinking.

Structure of the Book

This reimagination of higher education within this disruptive era of worsening climate injustice is structured around the links among knowledge, wealth, and power. Each chapter includes personal reflections based on my own experiences to illustrate the ideas being presented. Chapter 1 describes the transformative power of climate justice when considering the power and potential of higher education in an increasingly destabilized world. A simple triangular framework connecting knowledge, wealth, and power provides a novel way to consider trends in how higher education is currently engaging with climate injustices and how universities could shift to engage in new and different ways to better align with societal needs. Chapter 2 reviews critiques of current practices, trends, and priorities in higher education, exploring how universities leverage selective knowledge to concentrate wealth and power and reinforce climate injustices. This chapter situates universities as organizations rooted in patriarchy, racism, capitalism, and coloniality drawing from, and contributing to, the growing field of critical university studies. Acknowledging the legacy of extractive and exploitative practices within higher education is essential to moving beyond the narrow focus of promoting individual and institutional success and maximizing the financial status of the university and its students, faculty, staff, administrators, and alumni.

The next two chapters focus explicitly on knowledge, reimagining how the university disseminates knowledge (chapter 3) and creates knowledge (chapter 4). Chapter 3 proposes unlearning as a key concept for transformation because it provides a framework to imagine letting go of conventional ways that universities curate and organize their knowledge dissemination processes. Unlearning allows for a reimagining of what is possible with curriculum changes, pedagogical innovations, and epistemic pluralism. Unlearning can also be powerful because it encourages a humbling of higher education by reducing the arrogance and false sense of certainty that is often projected from academic experts claiming to know the best path forward. This chapter focuses on unlearning as a way to stop reinforcing narrow, distorted perceptions of how the world works. Unlearning of the simplistic way that mainstream neoliberal economics is taught in universities around the world is necessary to open up possibilities for envisioning alternative ways of structuring the economy, including, for example, a care-based economy (Lorek, Power, and Parker 2023) or a well-being economy (Fioramonti 2016; Chrysopoulou 2020). Unlearning is also required to make space for teaching relational knowledge and understanding about the planetary limits to the earth's systems.

To reimagine knowledge creation for climate justice, chapter 4 calls for prioritizing *exnovation* research, that is, research intended to improve understanding of the processes of phasing out or discontinuing materials, practices, products, or technologies. Rather than focusing research on innovation, *exnovation* research on fossil fuel phaseout and a plastic-free future are both urgent societal needs; yet this type of research is not well supported by academic funding sources. This chapter highlights the transformative potential of co-designing research and co-creating new knowledge with community partners outside of academia. It explores the dangers of the imbalanced research focus on technological fixes to the climate crisis (including geoengineering) when more research on social, economic, and cultural transformation is urgently needed.

Reimagining the role of universities in wealth distribution, chapter 5 explores alternative reparative and regenerative financial models for funding transformative climate justice universities. This chapter asks, what if wealthy universities like Harvard cultivated within their communities a different kind of generosity? Rather than strategically encouraging students, alumni, and others to make financial contributions to the university, what if they instead cultivated generosity based on the reciprocity that is central to so many indigenous cultures? If structured and funded differently, higher education institutions could be key players in redistributing and sharing wealth rather than the role many universities are currently playing in concentrating and hoarding wealth. Applying the "resist, reclaim, and restructure" framework adapted from the energy democracy movement, this chapter first focuses on resisting academic capitalism and the financialization of higher education. Then ideas for reclaiming higher education as a public good, rather than a private resource, are reviewed, and provocative proposals for restructuring the financial organization of universities to focus on wealth distribution and community wealth building are suggested. The potential benefits of restructuring higher education with a cooperative, worker-owned model are explored.

To reimagine the role of universities in distributing power and empowering local communities, chapter 6 explores new types of local community engagement with global solidarity. A paradigm shift centering communities and decentering universities is proposed, and examples of community-based university structures promoting civic engagement and transformation are presented. This chapter also introduces social footprint mapping, developed by Davarian Baldwin, a tool that every university could use to hold themselves accountable to the communities where they have impact and influence (Baldwin 2022).

The book concludes with a synthesis of the opportunities and resources for universities to redefine their purpose and values. Establishing a culture of care and accountability requires intentionally resisting dehumanizing frameworks that promote complacency to human suffering and disconnect us from each other and from the earth's regenerative power. The vision of a geographically distributed global network of universities emerges from the reconceptualization of higher education as critical social infrastructure that every community needs. In their 2023 book—*Not Too Late*—the feminist activists and storytellers Rebecca Solnit and Thelma Young Lutunatabua bring together the courageous voices of climate justice activists from around the world to change the climate story from despair to possibility. Higher education institutions have the potential to engage, embrace, and amplify these voices to advance global solidarity and climate justice.

CHAPTER 1

Transformative Climate Justice in Universities

As universities around the world grapple with how best to respond to the intersecting crises facing humanity, an inspiring event took place in Galway, Ireland, in early November 2023. Over one hundred academics and activists from universities throughout the island of Ireland convened at a hotel conference center for an all-day discussion focused on what higher education institutions should do about the planetary crisis. Co-organized by academic staff from the University of Galway (Sinéad Sheehan) and Queen's University Belfast (John Barry and Calum McGeown), faculty, staff, researchers, and students were joined with community activists for a nonhierarchical format, a world café–style event, to learn from one another's diverse perspectives and to develop action steps for this all-island network to collaborate on changing Irish universities.

Throughout the day, participants shared ideas for research and teaching in this time of accelerating ecological instability and collectively explored opportunities for connecting academic work with activism, outreach, and community engagement. A series of short provocations catalyzed lively conversations on the privileges, as well as the responsibilities, that come with academic freedom and the societal impact of universities. A collective sense of duty to change research norms and adapt learning expectations within academic culture was expressed, and the group co-developed a list of specific proposed action steps that included both simple near-term changes as well as longer-term systemic changes.

During my sabbatical in 2022, I spent time in Ireland as a visiting professor at Trinity College Dublin, where I connected with academics throughout the country by attending Irish academic conferences and giving presentations at universities throughout the island of Ireland. Although I was unable to attend this meeting in Galway, I heard from multiple colleagues and collaborators that this event was both inspiring and impactful.

Unlike more common top-down university strategic initiatives focused on sustainability and climate, this bottom-up convening included voluntary participation among a diverse set of academics and activists. This diverse group came together representing different universities and communities with shared concern and a cooperative commitment to transformative change. Participants traveled from throughout Ireland seeking a supportive environment to inform, co-create, and collaborate on how to adapt their university work to better respond to the worsening climate injustices of the world. The event provided a forum for some radical and transformative ideas to be collaboratively explored, many of which resonate with the ideas presented in this book. This convening in Galway created momentum for a series of subsequent meetings planned in each of Ireland's four provinces to further advance the collective work of this informal network committed to transforming higher education in Ireland and beyond.

During the event, co-organizer John Barry reported that the use of the Irish language emerged as a decolonizing and (re)indigenizing way for some participants to think about and act on the climate and ecological crises in Ireland. The day began with a welcome in the Irish language from the president of the University of Galway, Ciarán Ó hÓgartaigh, and then Professor Peadar Kirby, one of the academics invited to give a short intervention, made his remarks, also in Irish, explaining that when speaking in public he now always speaks in both Irish and English or only Irish. Subsequently, one group of participants chose to spend the whole day using the Irish language to discuss the changing role of higher education.

Although it may not be immediately apparent, language revival in places like Ireland, where colonizing powers have minimized the use of the indigenous language, is directly related to addressing the planetary crisis. The richness of the Irish language is closely tied to the natural landscape. Compared to the English language, which is now the dominant language in Ireland, the Irish language provides different ways for people to relate to the natural world. For example, in the Irish language thirty-two different words refer to a "field": the word *reidhlean* means a field used for games and dancing, *tuar* is the word for a field for cattle at night, and *cathairin* is the Irish word for a field that holds a fairy dwelling (Magan 2020). The loss of these nuanced words within the Irish language contributes to a loss of some of the ways that people in Ireland connect with the land (Cronin 2019).

Multiple aspects of this Irish network dedicated to transforming universities have inspired me. The convening in Galway demonstrates the widespread interest in acting on the powerful potential of universities to engage with transformative climate justice. One powerful structural element of this collective effort was the decentering of the institutional power of a single university. Although the University of Galway was the official host of this first meeting, the co-organizers and participants represented universities and communities throughout the island of Ireland and, therefore, no specific institution was the focus nor did any single university have the burden of implementation or the impulse to control the outcome. The collective focus on the higher education sector as a whole dissipated any sense of competition among institutions and facilitated a collaborative and cooperative approach.

The inclusive and participatory format of the day, and its intentional resistance to conventional academic hierarchies and modes of engaging, enabled everyone in attendance to engage and contribute in a meaningful way. The meeting was thoughtfully structured, with two experienced facilitators ensuring inclusive participation by all who were there. This inclusive emerging network in Ireland demonstrates the active engagement among academics and activists to leverage the power of higher education for transformative social change.

A Transformative Lens for Higher Education

Although higher education leaders often claim that their institutions contribute to addressing the world's biggest challenges, it is not always clear whether and how they do this. In this new era of polycrisis—a term popularized recently by historian Adam Tooze that describes intersecting, cascading, and self-reinforcing catastrophes (Tooze 2022)—it is increasingly acknowledged that large transformative systemic social change is urgently needed. Referencing the accelerating risks and the dire consequences of continued insufficient action, the United Nations secretary-general has called for a "quantum leap in climate action," confirming that we need "everything, everywhere, all at once" (Guterres 2023). Because climate change is interlinked with all aspects of society, responses to the climate crisis cannot be addressed in isolation of other grand challenges facing humanity, including war, militarization, mass migration, and mass extinctions. Although an interconnected transformative lens is urgently needed to reduce human suffering and stabilize humanity's future, universities are still largely organized within narrow and strict disciplinary departments based on a medieval way of organizing knowledge. These structures encourage isolating discrete problems rather than leveraging the interconnected links among different challenges for transformative systemic change.

Many have described how disciplinary boundaries, financial entanglements, and commodification within higher education institutions create structural barriers preventing universities from engaging with transformative social change (Kelly et al. 2023; Lachapelle et al. 2024; Lueddeke 2020; Urai and Kelly 2023). It is also widely acknowledged that investments in and incentives for social innovation toward systemic change are minimal compared to funding for technological innovation (van Damme 2021). Rather than trying to encourage and support research and teaching on systemic social change, most higher education institutions increasingly incentivize technological innovation, incremental change, and technical fixes to discrete, narrowly defined problems. Social innovation, which includes experimentation and exploration of mechanisms for economic, political, and cultural change, and can also include new modes of governance and alternative ways of working and living together (Pel et al. 2023), is often considered outside the realm of academic work. Transformative social innovation, which goes beyond traditional problem-solving by seeking new ways to co-create social value, promote equity, and address systemic marginalization, has not been a core mission of most higher education institutions.

The preference within higher education institutions for supporting technological innovation over social innovation results from many factors, including the male-dominated, patriarchal, colonial, and extractive culture of control that is perpetuated in universities. The disproportionate focus on technology also results from the greater likelihood of near-term financial rewards for those individuals and institutions who develop, adopt, and patent technological innovations. Societal benefits associated with researching alternative ways of structuring social enterprises or developing innovative approaches for deliberative democracy are not as tangible and discrete as developing a new app or gadget. Despite the large potential for transformative impact of investing in social innovation, quantifying the impact of exploring new organizational structures to promote local economies and generate community wealth may be more difficult to measure, and more challenging to find financial support for, than assessing the deployment of a new technology.

This dominant focus on technological innovation over social, political, and economic innovation has led to growing critiques of higher education and its limited capacity to engage with social and economic transformation (Patel 2021; Washburn 2005; Renner and Moore 2004; Fitzpatrick 2019; Baldwin 2021; Connell 2019). Prioritizing STEM (science, technology, engineering, and math) education is a reinforcing cycle because with reduced exposure to critical explorations of social, political, and economic structures, students accept, rather than challenge and engage with, our existing social systems that are not serving humanity well. The disproportionate focus on science and technology has contributed to a sense of complacency and disempowerment, which discourages engagement in the processes of social change. These critiques suggest that many aspects of higher education are, in fact, reinforcing the status quo and delaying transformation rather than facilitating and accelerating systemic social change.

In 2010, with my friend and collaborator Amanda Graham, a communication scholar and educator who was working at the Massachusetts Institute of Technology at the time, we described a fundamental paradox of higher education organizations and wrote that "they are institutions designed to teach but not to teach themselves" (Stephens and Graham 2010, 617). At the time, we were proposing an empirical research agenda for understanding the role of higher education in sustainability transitions, and we found the lack of self-reflexivity within universities surprising (Stephens and Graham 2010). Since then, I have appreciated many others making this same point; despite being organizations focused on learning, universities are not learning organizations (St. Clair 2020). Rather than accept this paradox, we can increase our ambition and expectation for universities and publicly acknowledge that *another university is possible.* Transforming universities is not only possible, it is essential for the societal and economic transformations that are urgently needed (Loorbach and Wittmayer 2024).

A transformative lens is needed within higher education because despite expanding vulnerabilities and growing precarity of people across the world, collective efforts to steer humanity toward a more healthy and stable future are increasingly ineffective. As corporate profits and billionaires' fortunes continue to balloon, social, political, and economic systems have been captured by what economist Marjorie Kelly calls "wealth supremacy"—the bias that institutionalizes infinite extraction for the wealthy reliant on expanding exploitation of people and the planet (Kelly 2023). Just as white supremacy characterizes the biased systemic and structural policies, practices, and priorities that privilege whiteness, wealth supremacy characterizes the biased systemic and structural policies, practices, and priorities that privilege wealth.

In an increasingly financialized society, financial markets and financial assets are often considered more important than the real economy of jobs, income, and spending. As governments around the world try to protect people from growing economic and ecological precarity, it is increasingly clear that continued concentration of wealth and power is hurting us all and thwarting needed transformative changes. Our disconnection from one another and the land that provides for us is making us physically and emotionally sick. But we are not stuck on this path, and we cannot and should not be complacent. Although transformative change may seem difficult to imagine, we are not locked in to these socially constructed constraints that continue to move us in the wrong direction.

Transformative changes in our economic, political, and educational systems are not only possible, they are becoming more and more likely as humanity faces the worsening polycrisis. To halt the damage and move humanity toward a more healthy, regenerative existence, it is increasingly clear that bigger systemic and structural changes are urgently needed. Processes of transformation are well underway, requiring collective envisioning and inclusive reimagining to allow ourselves to let go of the structures, practices, and ideas that are no longer serving us. The power and influence of universities around the world can be leveraged to facilitate and support these processes of social and economic change.

Climate Justice: Beyond Climate Isolationism

Recognizing the transformative potential of colleges and universities, the principles of *climate justice* provide a valuable framework for reimagining and implementing institutional change across the global higher education sector. Climate justice centers the inequitable vulnerabilities to climate impacts and focuses climate mitigation and adaptation efforts on redressing the injustices by reducing marginalization, exploitation, and oppression (Sultana 2022a). Climate justice assumes a commitment to transforming economic and political power, and climate justice can be applied at multiple scales: institutional, local, regional, national, and global. In centering power, transformative climate justice includes multiple kinds of justice, including *procedural climate justice*, which is about fairness in decision-making processes; *distributive climate justice*, which focuses on equity in the distribution of harms, benefits, and impacts; *recognition justice*, which refers to equitable representation, particularly of marginalized groups; and *intergenerational justice*, which focuses on fairness for future generations (Newell et al. 2021).

Climate justice is about paying attention, not just to the science of climate change, but also to the unequal and disproportionate impacts of climate change among different households, communities, and regions of the world (Stephens 2022). The earth's climate is changing rapidly because heat-trapping greenhouse gases, including carbon dioxide and methane, are accumulating in the atmosphere (IPCC 2022a). Fossil fuels are the biggest contributor to carbon emissions, and animal agriculture and industrial processes also play big roles. Biological systems absorb carbon from the atmosphere, but deforestation and land degradation are reducing these carbon sinks. With more heat-trapping gases in the atmosphere, the global average temperature is increasing and the earth's climate system is becoming increasingly destabilized, resulting in more intense and extreme weather events of all kinds, including heat waves, droughts, floods, and storms (IPCC 2022a). These disruptions cause wildfires, crop failures, lack of water, and sea level rise, impacts that are forcing people to leave their homes. These impacts are not felt equally around the world: in every region, it is marginalized communities and poorer households that are most vulnerable.

Climate justice is much broader and more inclusive than the narrow technocratic way that climate action is often presented (Stephens 2022). Beyond greenhouse gas emissions reductions and decarbonization, climate justice requires structural and systemic economic, political, and cultural change as well as solidarity and collective action. Climate justice includes what Dr. Jalonne White-Newsome, an environmental justice policy analysist, refers to as ADAPT-ing: Acknowledging the harm; Demanding accountability; Addressing racism, power, and privilege; Prioritizing equity; and Transforming systems (White-Newsome 2021).

The distinction between mainstream climate action and climate justice is critically important because many climate policies and technologies that are not justice centered have resulted in worsening economic and political inequities. For example, in many places public incentives for rooftop solar are accessible only to privileged households who can afford to invest in solar panels. While these incentives have expanded the deployment of solar panels, it is well-off families and privileged communities that now have more affordable, clean energy, while the energy burden (the percentage of household income used to pay for energy) in lower-income households has been increasing (Hernández and Bird 2010; Lennon 2017). Similarly, when the climate action plans and policies of universities focus only on incremental technology-based change and fail to consider equity and structural social and economic changes, the university exacerbates climate injustices by reinforcing inequitable access to cleaner technologies. This technology-focused approach further benefits those who are already privileged; all too often, technologies serve to protect and reinforce those privileges.

It is important to point out that climate justice is an approach more expansive and inclusive than the dominant method of presenting the climate crisis as a narrowly defined scientific problem in need of a technological solution. Climate action is often framed simply as greenhouse gas accounting and emissions reductions defined in relation to a scientific goal of stabilizing the global average temperature change at 1.5 or 2 degrees Celsius. This narrow technocratic way of approaching the climate crisis is what I call *climate isolationism* (Stephens 2022). I use the word *isolationism* to highlight the lack of integration. Most climate action focuses on inadequate policies, practices, and technologies that treat climate change as a discrete problem separate from our economic and political systems. Climate isolationism is a nontransformative approach that has proven to be ineffective in catalyzing the scale of social and economic change required for a climate-stable future. Climate isolationism is the dominant way that climate action has been framed in public dialogue (figure 1). In addition to being mainstream, climate isolationism can also be considered malestream, which is a feminist concept to describe when male scholars carry out research from a male perspective and then assume the findings can be applied to everyone (Guy-Evans 2023). While climate justice is based on feminist principles of challenging power dynamics, climate isolationism is perpetuated by a male-dominated technocratic view of the world that fails to consider power structures. Mary Robinson, climate justice advocate and former president of Ireland, acknowledged the ineffectiveness of the male-dominated climate isolationist approach when she characterized the disappointing UN climate negotiations in Glasgow in 2021 as "too male, too pale, and too stale" (Collins 2021).

The dangers of climate isolationism can be seen with the narrow focus on maintaining global average temperature change to less than 1.5 degrees. This goal is based on scientific analysis that suggests that climate instability increases dramatically once the global average temperature exceeds an increase of 1.5 degrees Celsius. This goal, which was formalized in 2015 during the international climate negotiations in Paris, has been helpful in providing a collective global target within which indi-

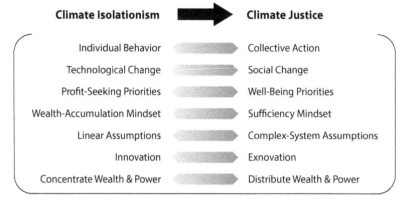

Figure 1 Climate Justice Paradigm Shift. To advance climate justice, a paradigm shift in higher education is needed: away from the characteristics of climate isolationism, prioritizing instead the principles of climate justice.

vidual countries can situate their emission reduction plans. But as it becomes less likely that the goal will be achieved, the target is being used to justify massive investments in technological innovation to develop carbon capture and carbon removal techniques rather than investing in social, economic, or political changes. The narrow technocratic view fails to imagine restrictions on fossil fuel supply, fails to integrate the climate impacts of war and militarization, and ignores all kinds of possibilities for investments in transformative social and economic change. As the global average temperature continues to increase in response to rising emissions and continued burning of fossil fuels, widespread despair, despondency, and fear are detracting from the social innovation, empowerment, and collective action that are needed for stability, peace, and justice.

Expanding beyond climate isolationism, climate justice focuses instead on investing in people and communities integrating feminist perspectives and insights (Sultana 2022a). Climate justice reveals and resists structures, policies, practices, and relationships that maintain injustices and perpetuate fossil fuel reliance. While mainstream climate policy focuses on market mechanisms, technological innovation, and individual behavior change, climate justice represents a much broader, holistic, and global approach to leveraging the urgency of the climate crisis to invest in reparative and regenerative futures.

This book's proposed new vision of climate justice universities requires a paradigm shift away from climate isolationism. Figure 1 illustrates the distinctions between the key characteristics of climate isolationism and the principles of climate justice. Climate justice requires a shift away from individual behavior toward collective action, from technological change to social change, from profit-seeking priorities to well-being priorities, from a wealth-accumulation mindset to a sufficiency mindset, from concentrating wealth and power to distributing wealth and power, from innovation to exnovation, and from linear assumptions to complex-system assumptions. These are not simple dichotomies; rather, they are different ends of a spectrum. To unleash the transformative power of climate justice, universities need to deprioritize the dominant emphasis of characteristics that reinforce a climate isolationism lens and instead prioritize the principles of climate justice.

Climate justice focuses on reducing systemic dehumanization, marginalization, exploitation, and oppression by enhancing equity and justice. By focusing on reducing inequities and disrupting systems and practices that are exacerbating disparate vulnerabilities, climate justice integrates the need for large and integrated investment in both climate mitigation and climate adaptation focused on the most vulnerable places and communities (Barrett 2013; Weinrub and Giancatarino 2015).

According to my colleague and collaborator, feminist anthropologist Frances Roberts-Gregory, climate justice requires "radical shifts in how we a) build and sustain relationships, b) manage uncertainty, disruption, grief, and shock, and c) redistribute wealth, opportunity, risk, and accountability" (Roberts-Gregory 2021).

To prioritize climate justice initiatives that move beyond climate isolation, higher education institutions need to diversify and expand beyond conventional academic knowledge frameworks. Universities also have to move beyond hierarchical ways of working to strengthen collaborative relationships with broader networks outside of the university. A paradigm shift is needed for universities to embrace climate justice principles. Higher education can move beyond siloed disciplines and conventional knowledge hierarchies to instead integrate and value other kinds of expertise, experiences, and perspectives. Reimagined alternative teaching and learning opportunities are explored in chapter 3, knowledge co-creation is discussed in chapter 4, and new ways of centering community relationships are proposed in chapter 6.

Transparency and accountability are key to climate justice and also central to the idea of climate justice universities. Consistent acknowledgment of harm caused by both current practices and legacy systems is essential for dismantling and disrupting entrenched exploitative and extractive structures within universities. Denial, defensiveness, and covering up negative impacts are contrary to climate justice. Advancing climate justice, therefore, provides an innovative framework for universities to broaden their societal role, become transparent and accountable, and expand beyond a narrow focus on individual success of their students, their researchers, and their individual institutions.

The current focus in higher education on competitive assessments to define "success" is reinforcing exploitative societal assumptions about education and research; in order for some students, faculty, or institutions to rise to the top of academic ranking systems, other individuals and institutions must move down to the bottom. This pervasive competitiveness relies on systemic exclusion of some people—particularly systematically marginalized and nontraditional people; it also impedes cooperation and collaboration. A commitment to climate justice challenges these dominant paradigms within higher education and enables a transformative lens. By refocusing on collective action for the common good, climate justice incentivizes education and research that empowers individuals and institutions to embrace and promote nonexploitative practices, policies, and priorities.

In this era of polycrisis, the climate crisis is just one among many crises. Prioritizing climate justice as a guiding framework for transformative change in higher education is warranted not because the climate crisis is the most important challenge, but because it is exacerbating all other challenges. In fact, US-based climate justice activist and human rights lawyer Colette Pichon Battle has explained that "climate change is not the problem . . . but the most horrible symptom" (Battle 2020) of a capitalist, extractive economic system. Proposing climate justice as a guiding framework for universities makes sense because of its all-encompassing, inclusive urgency. The climate crisis is different from all other crises because it involves a destabilization of every aspect of human systems and earth's systems—and the pace of change is accelerating at an alarming rate, demanding timely responses. The inevitable global changes that are already causing devastation, suffering, and forced migration around the world require coordinated, collective, care-based responses.

Climate Obstruction in Academia

A critically important aspect of embracing transformative climate justice in higher education requires resisting *climate obstruction* within academia. Climate obstruction refers to any action aimed at stalling climate policy or climate action (CSSN 2021). Climate obstruction includes outright denial of the climate crisis and intentional efforts to delay and distract from climate action (Ekberg et al. 2022). Expanding research on the scale and scope of climate obstruction suggests that universities continue to be leveraged by powerful economic interests to legitimize climate obstruction narratives, tactics, and analysis (Morris and Jacquet 2024; Hiltner et al. 2024; Brulle and Dunlap 2023).

For decades, the science has been clear and unequivocal—climate instability is resulting in more intense and frequent extreme weather events (IPCC 2021); and climate disruptions are causing greater devastation and suffering in communities already marginalized (Ackerman and Stanton 2008; Denton 2002; Kane and Shogren 2000; IPCC 2022a). The global average temperature of the earth has been steadily increasing, and fossil fuel burning, the largest contributor to accelerating climate change, continues to expand (IPCC 2021). Since the 1980s and 1990s, when climate change emerged as a future threat, the clarity of the scientific and empirical evidence for rapidly phasing out fossil fuels and investing to reduce climate vulnerable communities has grown stronger (Ghosh 2017; IPCC 2014, 2007). Each year, more of the dire projections about the worsening devastating impacts of unabated climate change—including catastrophic hurricanes, forced migration from sea level rise and flooding, and deadly heat waves—come true.

Despite the clarity of the science, the fossil fuel industry and others profiting from continued fossil fuel reliance have invested in a complex network of climate obstruction efforts since the 1980s (Ekberg et al. 2022; Carroll 2021; Oreskes and Conway 2010). Universities and university researchers have played a central role in climate obstruction, with a network of fossil fuel companies and their allies leveraging academia to delegitimize the science of climate change and advocate against climate policy and climate action (Wilson and Kamola 2021; Brulle and Dunlap 2023). Coordinated resistance to climate policy is embedded within a much longer history of industry investing in US-based universities to promote the ideology of market fundamentalism, the belief that unregulated free markets are essential for freedom and prosperity (Oreskes and Conway 2023). In their 2023 book, *The Big Myth: How American Business Taught Us to Loathe the Government and Love the Free Market*, historians Naomi Oreskes and Erik Conway trace over one hundred years of industry strategists influencing university professors, university research, university curriculum, and even university textbooks as part of a coordinated effort to strengthen the power of the private sector by weakening the power and influence of the government.

Those profiting from fossil fuel extraction are funding an extensive network of climate obstruction (Supran, Rahmstorf, and Oreskes 2023; Dunlap and Brulle 2020), including a steady stream of funding to universities (Ladd 2020). In addition to funding, fossil fuel companies have embedded themselves within many universities by incentivizing executives to serve on university boards and by engaging directly with research initiatives and research partnerships (Carroll, Graham, and Zunker 2018). Fossil fuel companies, as well as many governments and politicians reliant on financing from fossil fuel profits, resist plans to phase out or even phase down fossil fuel supply. Instead, they continue to invest strategically to appear committed to climate action while doing whatever they can to maintain fossil fuel reliance to sustain their corporate profits (Li, Trencher, and Asuka 2022; Newell, van Asselt, and Daley 2022; Si et al. 2023).

Fossil fuels are among the earth's most versatile resources, used not only for energy but also in many industrial processes, including fertilizer and plastic manufacturing. While extracting, processing, and burning fossil fuels has enabled many aspects of contemporary society, the extensive and transboundary harms to people and the planet have been dismissed and disregarded because of the financial gains to the privileged few (Healy, Stephens, and Malin 2019a). Those advocating for continuing fossil fuel exploration argue that neither fossil fuel phaseout nor fossil fuel phasedown is necessary because the emissions from fossil fuels can be addressed through technologies that capture, remove, and store carbon. But this climate-isolationist technological approach dismisses the many other reasons—beyond the climate crisis—to phase out fossil fuels, including devastating health impacts and ecological destruction (Healy, Stephens, and Malin 2019b). Carbon capture and removal technologies are also extremely expensive, energy intensive, and not available at scale (Stephens 2014; IPCC 2005).

International climate policy has been co-opted by fossil fuel interests committed to dismissing climate justice and perpetuating climate isolationism to maintain the status quo. The effectiveness of this strategy was demonstrated when the managing director and CEO of the Abu Dhabi National Oil Company served as president of the December 2023 United Nations climate conference (COP-28). While climate activists around the world have been calling for years to "leave it in the ground," fossil fuel suppliers are doubling down on their strategic investments to delay any policies or actions to address climate that would disrupt their plans to continue profiting from fossil fuel expansion (Newell and Simms 2020).

A key component of this strategic climate obstructionism includes continued investments in higher education to legitimize delay by promoting the promise of nontransformative technologies and constraining inquiry on social change (Leonard 2019; Graham 2020). Research confirms that fossil fuel companies benefit from university partnerships by purchasing academic credibility and public trust of universities and using research outputs to lobby policymakers (Gray and Carroll 2018). A 2023 report revealed that fossil fuel companies donated more than $700 million in research funding to universities in the United States from 2010 to 2020 (Data for Progress 2023), and a 2022 study of twenty-six academic energy research centers in the United Kingdom, United States, and Canada confirmed that those funded by fossil fuel interests wrote more positively about natural gas than renewable energy (Almond, Du, and Papp 2022).

Many climate and energy research centers and degree programs at universities around the world are funded by industry partners who are then invited to engage and contribute to shaping the university's initiatives (Hiltner et al. 2024; Corderoy et al. 2023). Universities have become influential and strategic sites of climate obstruction because corporate interests have leveraged science and scientists to delay action by reinforcing what epidemiologist David Michaels has coined "manufactured uncertainty" (Michaels 2020). The influential societal role of higher education institutions has been clearly recognized and leveraged by fossil fuel interests to thwart climate action, perpetuate fossil fuel reliance, and sustain the status quo by prioritizing the development of future technologies to distract from fossil fuel phaseout.

In human societies, higher education institutions are central nodes within the complex interconnections among knowledge, wealth, and power (figure 2). As organizations that exert

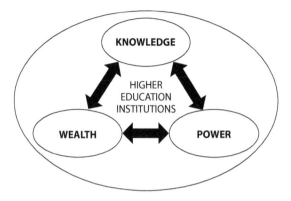

Figure 2 Knowledge, Wealth, and Power. Higher education institutions are a central part of complex interconnections among knowledge, wealth, and power. *Source:* Adapted from Sokol (2013).

power by disseminating and generating certain kinds of knowledge, many higher education institutions rely on and reinforce wealth accumulation. In response to financial pressure, universities around the world are increasingly reinforcing economic and political processes that intensify inequities and disparities. With the global concentration of wealth and power among a small number of billionaires, higher education institutions have become vulnerable to the whims of the super-rich; the power of knowledge has been co-opted to justify and perpetuate unjust exploitation and devastating extraction.

Decades of ineffective and inadequate climate policy demonstrates how attempts by powerful actors to delay transformative climate justice maintain the stability of current systems. The effectiveness of these delay strategies has contributed to our collective inability so far to redesign and restructure our economic and political systems. Humanity is in a much more difficult place now than we were in the 1970s and 1980s, when our collective vulnerabilities to the climate crisis and ecological devastation first became apparent. Within universities, attempts to distract from or minimize attention to climate injustices and ecological collapse can also be viewed as efforts to maintain the stability of current systems. A collective complacency within higher education has been nurtured by the legacy of narrow disciplinary boundaries that constrain our collective imaginations about alternative economic and political futures (discussed more in chapter 3) and about alternative incentive structures that are currently constraining research (the focus of chapter 4). This complacency is reinforced by financialized university structures that promote individual and institutional success instead of collective well-being (explored in chapter 5) and by the disconnection from societal needs that comes with centering the university instead of centering communities (discussed in chapter 6).

Universities are institutions responsive to and embedded within our economic, political, and cultural systems, so despite their focus on truth, knowledge, science, and open inquiry, universities have not been shielded from climate obstruction. Rather, higher education institutions have been and continue to be heavily influenced by the same powerful forces that are influencing other sectors of society. There is growing awareness that higher education institutions are in fact cogs in the wheel of obstruction, delay, and deception—not just on the climate crisis but also contributing to economic injustice, structural racism, and gender oppression. Deep and genuine engagement with these systemic and structural issues requires courage and a difficult, risky self-reflection that is not generally rewarded or encouraged within university systems.

Preventing continued climate obstruction in academia requires transformative change in how higher education is funded (more on reimagining alternative models of finance for universities in chapter 5), as well as democratization of how universities are run and managed (McGeown and Barry 2023). Corporatized management of universities and the disproportionate influence of wealthy and powerful individuals and organizations on knowledge production and knowledge dissemination in higher education are incompatible with transformative climate justice. Until and unless billionaires and corporate interests are no longer able to use their extreme wealth to exert power over universities by incentivizing them to embrace their priorities, climate obstruction in academia will continue. To leverage the transformative power of higher education for climate justice, therefore, a rebalancing of the links among knowledge wealth and power is needed.

Political Power in Higher Education

Throughout human history, powerful people and institutions, including political leaders and religious institutions, have leveraged their power and influence of the university to advance their priorities. When authoritarian governments, oppressive regimes, and conservative politicians in different parts of the world feel threatened by the freedom of inquiry within higher education, many try to control the perspectives being taught and discussed in universities. While fossil fuel companies have provided funding to universities to have influence, withholding funding (or threatening to withhold funding) is another way to restrict areas of inquiry in universities. Other forms of intimidation include discrediting and public shaming, firing academics who voice specific political views, and banning certain topics or perspectives from the university curriculum.

An example of political intimidation of university leaders focused on attempting to discredit and publicly shame took place in December 2023 in the United States Congress. In the midst of public controversy related to student protests on campus advocating for peace and liberation for the Palestinian people after the horrific October 7 Hamas attacks on Israel, the presidents of three prominent universities (Harvard, MIT, and the University of Pennsylvania) were called to testify to the US House Committee on Education and the Workforce at a hearing titled "Holding Campus Leaders Accountable and Confronting Antisemitism" (Helmore 2023). Before the hearing, the chair of the committee, Virginia Foxx, a Republican from North Carolina, said, "By holding this hearing, we are shining the spotlight on these campus leaders and demanding they take the appropriate action to stand strong against antisemitism" (Helmore 2023).

For an example of firing academics who voice specific politic views, we can look to Turkey in 2016. Hundreds of professors were abruptly fired from their university jobs, and many face ongoing criminal persecution after signing a peace petition demanding an end to fighting with Kurdistan (Scholars at Risk Network 2023). Over one thousand scholars from eighty-nine Turkish universities signed the peace petition, and the government's attempts to silence them and punish some with imprisonment has led to a culture of fear that has left a devastating impact and done irreparable harm to Turkey's higher education sector and its democracy.

An example of political influence by banning specific ideas from the curriculum can be seen in the United States in 2023, when the governor of Florida, Ron DeSantis (a presidential candidate in the 2024 election), implemented efforts to reform Florida's state university system by abolishing programs designed to promote diversity, equity, and inclusion (DEI). As part of this initiative, he proposed dismantling the faculty tenure system that protects freedom of inquiry and declared that all students be required to take general education courses in history and philosophy that have shaped Western civilization. Under his leadership, Florida has enacted multiple laws restricting what can be taught with regard to race, sexuality, and gender, and DeSantis wants to prioritize "students with degrees that lead to high-wage jobs, not degrees designed to further a political agenda" (Contreras 2023). DeSantis is a leading politician claiming that higher education needs to be reformed because universities and colleges around the country are promoting "woke" culture, a term that has been captured by conservatives to describe progressive political priorities in a derogatory way. In this context, woke culture includes those who resist social injustices of police brutality against Black communities as well as those promoting racial equity and gay and trans rights. These recent examples from Turkey and the United States demonstrate how political power is leveraged to control higher education and how authoritative leaders feel threatened by efforts to foster a diverse, free, peaceful, pluralistic society within institutions of higher education. This politicization should not discourage or dissuade us, but rather inspire us to fight harder for leveraging the liberatory power of universities for justice and equality.

Politicization of higher education is not new. In fact, it has been a steady part of the history of universities: it was political, spiritual, and religious priorities that inspired the establishment of the earliest universities over one thousand years ago. The world's first university, the University of al-Quaraouiyine in Morocco, was founded in 859 during the Golden Age of Islam by Fatima al-Fihri, a visionary, wealthy woman who created a leading spiritual and educational center of the Muslim world when she used her inheritance to form a large mosque with an associated school (Tasci 2020). The University of Bologna in Italy began as a school of civic law in 1088, responding to conflicts between the Catholic Church and the Roman Empire (Gray 2001). The charter to establish the first university in Ireland, Trinity College Dublin, was granted in 1592 by Queen Elizabeth I as a strategic part of the colonizing Tudor monarchy extending its religious and political authority over Ireland. In the United States, Canada, and Australia, the founding of universities was a strategic part of settler colonialism, which included the violent claiming of land ownership for university buildings, prioritizing European knowledge and language, and excluding indigenous people (Stein 2022). The justification for establishing universities includes a common acknowledgement of the power of knowledge and its links to wealth accumulation.

Although some narrowly focused twenty-first-century capitalist free-market thinkers try to promote the idea that there is a kind of apolitical neutrality of knowledge and education, it is impossible for universities to hide behind a façade of apoliticality. Despite claims that the protection of academic freedom requires universities to be institutionally neutral, higher education has always been inherently political. No decision, research project, university course, or module is neutral or unbiased. Every program, idea, policy, or practice in higher education has political implications and was developed within a particular political and economic context. So rather than striving for an unattainable apolitical environment in higher education, universities have a responsibility to promote diversity and plurality of political thought with a directionality toward equity, justice, planetary health, and well-being—that is, climate justice.

Those who argue that universities should not get involved in political struggles are failing to acknowledge the multiple interconnected, inevitable, and undeniable links among knowledge, power, and wealth that reinforce certain ideologies (figure 2). So rather than resisting the politicization of universities, the opportunity ahead is to leverage the power of the university and its influence on wealth and knowledge to advance a more equitable, pluralistic, peaceful, and stable society.

Just as Ibram X. Kendi, leading scholar of race and discriminatory politics in America and author of the book *How to Be an Antiracist* (2019), says it is impossible to be neutral on racism and racial justice, it is impossible to be neutral on climate justice, social justice, and economic justice. The impetus for individuals and institutions to ignore the injustices of the climate crisis and be complicit in reinforcing prevailing mainstream thinking is strong, but there is also growing awareness of the opportunities to resist that mainstream thinking, to reclaim the potential of collective action for the common good, and to restructure society. As the climate crisis worsens and the injustices become more apparent even to those with concentrated wealth and power, the role of universities as critical nodes of inquiry and social change will continue to be scrutinized and leveraged (Alexander 2023).

In the 2019 book *A Planet to Win: Why We Need a Green New Deal* Kate Aronoff and coauthors make the case that "all politics is climate politics." With this statement, these scholars point out that everything is now impacted by the climate crisis due to the growing scale of climate disruptions. No decision or investment can be made without impacting or being impacted by intensifying climate chaos. If all politics is now climate politics, and higher education is inherently political, it is time to consider climate justice as a guiding principle for reimagining the role of higher education in society.

The Paradox of Hegemonic Power

This transformative reimagining of higher education engages directly with power struggles and the tensions between resisting versus reinforcing dominant, hegemonic policies, practices, and priorities. Hegemony refers to the processes that shape social, political, and cultural structures that allow a dominant group to influence subordinate groups by controlling the ideas, values, and cultural norms of society. This term was first developed and popularized by the Italian Marxist philosopher and social theorist Antonio Gramsci in the early twentieth century (Gramsci 2011) and has become an important concept widely adopted and developed in multiple fields, including sociology, political science, cultural studies, and international relations. Gramsci suggested that the ruling class achieves hegemony by establishing consent, not only by coercion (although this is important), but also by maintaining cultural and intellectual leadership. Hegemony has become a key analytical tool for understanding power dynamics, injustices, domination, and resistance (Faber et al. 2017; Levy 2005). The concept of hegemony is also critically important for understanding humanity's inability to respond to the existential threat of climate change.

Resistance to hegemonic power is constantly occurring in both subtle and obvious ways, resulting in a continual redefining and adaptation of the narratives that sustain the dominant power. To reproduce and reinforce the hegemonic interests and worldview, hegemonic practices are dynamic and ever-changing. In global political economy, the United States' hegemony, the dominant position and influence exerted by the United States over global politics, economics, and culture, is a prominent example. The hegemony of patriarchy, the dominant position and influence of patriarchal systems and structures whereby power and privilege are predominantly held by men, while women and gender nonconforming individuals are marginalized and disadvantaged, is another pervasive form of hegemony.

Higher education institutions around the world are engaged in both reinforcing and resisting hegemonic power, including

US hegemony and the hegemony of patriarchy. The US has a strong presence in global higher education, with many prestigious universities often setting standards in academic disciplines, research methods, and publishing norms. The dominance and prevalence of US-centric perspectives, theories, and knowledge frameworks reinforce the influence of US academia. At the same time, many university programs, students, and scholars are involved in collaboratively resisting US hegemony by revealing the devasting global consequences and economic injustices perpetuated by US imperialism. With regard to the hegemony of patriarchy, men hold the majority of leadership positions in higher education institutions around the world, and universities perpetuate a male-dominant culture of learning and research. At the same time, universities around the world are investing in multiple different programs and initiatives to promote and support women and nonbinary people.

These examples highlight the complex plurality of higher education institutions; universities are not monolithic but rather encompass a diverse range of programs, people, and perspectives. These examples also show that resisting hegemonic power involves paradoxes and inconsistencies because when an individual, organization, or institution takes action to resist hegemony they are inevitably also taking other actions that are reinforcing hegemony.

Adopting a transformative lens to consider the future of universities requires embracing the paradox of hegemonic resistance. We can expect simultaneous resistance and reinforcement of dominant practices, ideas, and norms, and we can try to accept the imperfection and discomfort of resistance. Resisting hegemonic power and advancing transformative change requires us to reorient ourselves and try to navigate in a different direction than the one we are currently moving toward. Transformative change is disruptive and requires learning new ways of thinking and doing while unlearning dominant paradigms that are no longer serving us.

Unlearning, the process of intentionally letting go of previously acquired knowledge, beliefs, assumptions, and behaviors, is a critical part of engaging with transformation and resisting hegemony. Unlearning is not easy or comfortable because it requires self-reflection and the questioning of our biases, prejudices, and assumptions (see chapter 3). In addition to learning and unlearning, transformative change and resisting hegemonic power also requires attention to both innovation and exnovation. Almost all universities around the world promote innovation, which includes the creation, development, and implementation of new ideas, technologies, processes, and services. But higher education institutions have yet to devote sufficient attention to exnovation, the deliberate process of phasing out and intentionally abandoning existing technologies, practices, and products that are deemed obsolete, inefficient, or harmful (Davidson 2019) (see chapter 4).

Reclaiming Higher Education Finance for the Public Good

Universities are generally perceived as organizations focused on education, learning, and knowledge creation. But increasingly, higher education leaders are also focusing on growth, expansion, and accumulation of resources. This "financialization" of higher education, which is happening all over the world but seems to be most extreme in the United States, has resulted in strategic decision-making in many universities that is influenced by private sector opportunities rather than by urgent societal needs of marginalized people and communities. When "successful" universities define themselves based on financial growth in their endowments and expanded real estate development, they minimize the value of their workforce, they disregard the needs of local and regional communities, and they limit intellectual inquiry by prioritizing educational outcomes that are financially lucrative. With universities increasingly operating more like private businesses serving wealthy customers rather than public entities serving the public good, higher education institutions intensify their role in segregating society by worsening economic inequities and disparities (McGeown and Barry 2023).

Although institutions of higher education are often presented as places that promote economic mobility (a university education can open up economic opportunities to low-income students), recent trends toward financialization of higher education results in universities actively contributing to the concentration of wealth and power among privileged elites. Universities increasingly cater to private sector partnerships and corporate interests rather than prioritizing education and research for the common good.

These processes of exclusion and hierarchy that are central to higher education institutions in most places around the world are reducing public support for, and perceived value of, investing public money in higher education. When higher education is a central node in the concentration of wealth and power among individuals and organizations who are already privileged, the priorities of universities are increasingly aligned with the priorities of the rich and powerful rather than with those of the broader public. Many of the richest individuals and most influential corporations maintain strong affiliations with higher education institutions. These university relationships are increasingly leveraged to promote their interests by distracting from large-scale transformative social change and resisting investments in a transformative response to the climate crisis and other intersecting crises (Kenner 2019; Stephens 2020). If higher education institutions were less exclusive, they would be more relevant to more people and would be better able to serve the common good. Reorienting universities to focus on climate justice and the goal of reducing climate vulnerabilities of communities around the world provides a guiding framework to reimagine, reinvigorate, and restructure higher education.

Doing Good? Good for Whom?

The societal expectation is that colleges and universities "do good." This widespread assumption that higher education provides a public service for the common good justifies multiple kinds of publicly funded financial support that governments provide to higher education institutions (Baldwin 2021). But the details of how universities "do good" and the distribution of who benefits are not well defined. In the business world, doing good while simultaneously maximizing profits is considered virtuous and rare. The designation of "B Corporation" was introduced in 2006 in the United States in an effort to incentivize and expand the vision of business as a force for good. Companies that are B Corps go beyond prioritizing the economic "bottom line" and integrate social and environmental considerations into their decision-making.

How universities do good is a question that lacks specificity and accountability. If every individual, every organization, and every university is committed to doing good, how do we reconcile all the negative trends in terms of global health, economic equity, and climate instability? If universities were "doing good" in their local communities by contributing to social and economic justice, a city like Boston with the highest density of universities in the world would be a world-class example of equity and justice (Boyle and Stephens 2022). But in fact, Boston is among the most unequal and racially divided cities in the United States. Clearly, the intention and declaration of "doing good" is insufficient for prioritizing initiatives and organizations that promote social and economic justice.

In the powerful 2019 book *The Good University: What Universities Actually Do and Why It's Time for Radical Change*, Raewyn Connell provides an expansive critique of how universities have been exacerbating economic inequities locally and globally. Connell reveals the unequal ways universities treat their own employees. She points out that by contracting out and creating conditions for high staff turnover, universities are devaluing the interconnected web of respected human relationships that is central to healthy communities oriented toward a common mission. With broad geographical scope, she explains the dangers of the power imbalance that the academic hegemony of the "Global North" creates, describing how the hoarding of funds, researchers, conferences, and journals reinforces imperial legacies perpetuating classism, sexism, and racism.

Expanding on Connell's call for radical change toward a "good" university, in this book I propose a paradigm shift toward conceiving of higher education institutions as critically important organizations for social change. I propose that not only are universities an underleveraged resource in society but that they also currently fail to be self-reflective and self-critical in assessing their power and influence, and in acknowledging the harm they may be causing. Given the capacity in higher education for data collection, analysis, and theoretical development, one might expect higher education institutions to be constantly applying their qualitative and quantitative research skills to evaluate their own societal impact; paradoxically, universities have been surprisingly unreflective when it comes to analyzing their impact in the world (Eaton 2022). While financial pressures have resulted in some universities investing heavily in data analysis for student recruitment and admissions, the same level of investment has not been made in assessing their own social and economic impact.

In this destabilizing time, it is dangerous to blindly accept university claims of "doing good" without interrogating the breadth of impacts that they are having on different communities and diverse aspects of society (McGeown and Barry 2023). The legacy of harm resulting from university activities needs to be openly acknowledged, and the potential for contributing to current and future harm also must be incorporated into higher education practices. Reimagining climate justice universities provides a framework to redefine and reinvent what it means for higher education to "do good" in the world.

A Paradigm Shift to Reduce Human Vulnerabilities

Communities in every region of the world are facing multiple crises that are intensified by climate disruptions and extremes. As more people are experiencing economic precarity, housing and food insecurity, growing inequities in access to quality health care and education, as well as intense geopolitical violence, the expansion of human suffering and vulnerability is striking and unprecedented. Intertwined with human crises are the ecological crises of species extinction and environmental degradation of land and water. In 2015, the United Nations developed seventeen Sustainable Development Goals (SDGs), which have attempted to focus global governance on the intersecting ambitions of eradicating poverty and hunger; ensuring health, education, and clean infrastructure for all; and reducing inequalities. In addition to critique about the contradictory, growth-based, neoliberal nature of the SDGs that are not based on human rights (Arora-Jonsson 2023), recent analysis shows that most of the metrics that assess progress toward achieving the SDGs have been moving in the wrong direction: rather than moving closer to eradicating poverty by 2030, the world is further away from that goal than ever before (Nishitani et al. 2021).

It is becoming increasingly clear that worsening human precarity, ecological collapse, the rise of authoritarian leaders, expanding militarization, the climate crisis, and the biodiversity crisis are all symptoms of a larger systemic problem; the exploitative capitalist institutions and financial structures are concentrating—rather than distributing—wealth, power, and resources. Human precarity and ecological devastation are intensifying around the world because current systems and structures incentivize extractive, competitive, and individualistic practices rather than regenerative, collaborative, and collective practices. In response to this downward spiral of worsening human suffering and ecological instability, global calls for societal transformation are growing stronger.

It is within this context that attention to, and concern with, the societal role of higher education in advancing transformative social change is growing. In 2022, the United Nations Education, Scientific and Cultural Organization (UNESCO) released a report focused on transforming higher education for global sustainability. The report highlights that although higher education institutions are uniquely positioned to coordinate and accelerate social, economic, and environmental transformation, systemic barriers are inhibiting this potential. Higher education institutions are constraining diverse ways of thinking and knowing within narrow legacy disciplinary boundaries that are preventing new approaches to producing and circulating knowledge (UNESCO 2022). The deeply entrenched models of university structures are limiting societal engagement, reducing the relevance of academic initiatives, and constraining societal impact (UNESCO 2022).

In the same year, UNESCO also released a draft roadmap

for "reinventing higher education" as a working document before the May 2022 third World Higher Education Conference. This roadmap acknowledges the inadequacy of higher education's contributions to sustainability and provides six key principles for building a new social contract for higher education. These include (1) inclusion, equity, and pluralism; (2) academic freedom and participation of all stakeholders; (3) inquiry, critical thinking, and creativity; (4) integrity and ethics; (5) commitment to sustainability and social responsibility; and (6) excellence through cooperation rather than competition. This declaration that the principles of inclusion, equity, pluralism, and cooperation are essential suggests that the current reliance within higher education on exclusion, inequity, elitism, and competition must be disrupted, resisted, and adapted. What is needed is a restructuring of how higher education is organized and supported.

In response to this need, the paradigm shift I propose includes acknowledging that higher education institutions have more democratic, transparent, and accountable power and influence in society than is commonly realized, and universities are currently underleveraged in how they contribute to the needs of humanity. Universities do much more than prepare individual students for successful, engaged, fulfilling lives and generate and disseminate new knowledge and technologies. Universities are also civic spaces and key sites of political contest, public debate, and civic deliberation, often shaping, reinforcing, or minimizing different cultural narratives and policy agendas. In diverse, pluralistic societies of the twenty-first century that face ever-increasing climate chaos, the power and influence of higher education must be carefully and intentionally leveraged in a reimagined way to prevent that power and influence from being co-opted by powerful elites to reinforce and sustain the exploitative systems that are worsening climate instability and injustices.

The idea of climate justice universities offers hope about the future of human societies and guides action toward a better future for all. The paradigm shift toward climate justice universities assumes that the teaching, learning, knowledge cocreation, and knowledge sharing that are at the core of higher education institutions could and should advance—rather than impede—the emergence of a more just, stable, equitable, and healthy future.

CHAPTER 2

Injustices of Higher Education

The patriarchal, exploitative structure of universities was revealed to me in an undeniable way with a life-altering sexual harassment experience I had at Harvard University over twenty years ago. I was a 27-year-old mother of two toddlers working at Harvard as a postdoctoral researcher when an internationally esteemed professor invited me to his office and tried to initiate a sexual interaction with me. This man, who I soon learned was a sexual predator, was not my direct supervisor, but he was part of a network of sustainability scholars within Harvard that I worked with. At a December holiday party on campus, he feigned interest in my research and then suggested that we schedule a time to meet so that I could tell him more about my project. He was well known and well respected, and I knew that making connections with individual professors at Harvard could be helpful to me in advancing my career, so after the party I enthusiastically followed up by email to arrange a time to meet; he replied suggesting that I come to his office at a specific time the following week.

I naively entered his office with excitement to share my research interests. As we sat next to each other on adjacent armchairs in his office and I began talking, I soon realized that he was not listening to what I was saying. While I was explaining my project, he began touching my knee, and he reached over to touch my cheek. He then explained to me that he had stayed up all night working on a grant proposal, and he was hoping that I could help him relax. As my brain shifted quickly to fight-or-flight mode and I realized what was happening, I stood up abruptly and tried to leave. He encouraged me to stay and touched me again. I successfully extracted myself from his office after some awkward conversation in which I tried to remain calm and polite. Before leaving the room, I ended up having to forcefully push him away to resist him pulling me toward him.

Although this physical altercation took only a few minutes, the impact of this interaction was life altering in multiple ways. My innocent perception of the university community as a safe nurturing environment was shattered, and I quickly and intuitively recognized the danger and violence of this exploitative situation. I felt danger not just for myself as a young woman and an emerging scholar, but I also felt danger for the others before me who I knew must have also experienced this professor's predatory behavior. The casual, smooth, and undramatic way he touched me and tried to draw me toward him made it clear that this was not unusual for him; he acted as if this was normal, typical, commonplace. He made me feel like I was the one being unnecessarily dramatic with my abrupt departure.

In the weeks and months that followed, with support from my family and friends, I cautiously navigated Harvard's sexual harassment reporting system. Throughout this process, I learned how the exploitative violence of that encounter was not universally acknowledged as dangerous, and the university procedures designed to respond were secretive and ineffective in holding him accountable. I submitted a formal complaint despite warnings of possible implications for me and my career. My complaint triggered an internal investigation, which resulted in a classic case of "he said, she said." The professor denied any inappropriate behavior, and because there were no witnesses or evidence, I was treated with suspicion.

The investigation explored whether I had any reason to make up the allegations. My credibility was scrutinized as the investigators interviewed my colleagues to assess my motivations for filing the complaint.

After months of uncertainty and a series of stressful meetings, I was called into an intimidating meeting with the dean, who told me the investigation was complete, the administrators believed my side of the story, and that the professor would be subject to disciplinary action. The dean also explained to me that due to privacy laws, I would not be told about the details of the specific disciplinary action that Harvard was taking against the professor because his professional record was confidential. That man remained in his post at Harvard until his death many years later. I carefully navigated my career, making multiple professional decisions along the way to strategically avoid future interactions with him.

I share this personal story to highlight the impact of sexism and sexist exploitation in academia and to demonstrate how patriarchal institutions facilitate and protect predatory and exploitative behavior, thereby reinforcing gender-based oppression. Upholding patriarchal structures not only creates dangerous, unwelcoming working conditions for many, but it also constrains diversity of ideas and perspectives and thus perpetuates injustices, including climate injustices. The famous quote by Black feminist activist Audre Lorde—"The master's tools will never dismantle the master's house"—reminds us how institutional norms and practices are constantly reinforcing the institution.

Collective reimagining of academic structures that break down, rather than reinforce, problematic hierarchical power structures first requires a shared acknowledgement of what is wrong with current academic systems. Before we can reclaim and restructure a new reimagined role for universities in transforming society toward a more equitable, healthy, and climate-stable future, we need to understand how current university structures leverage links among knowledge, wealth, and power to constrain social justice and restrict transformative change.

Universities are organizations rooted in patriarchy, racism, capitalism, and coloniality. Many current practices, policies, and priorities of higher education institutions continue to reinforce the legacy of these hierarchal systems in which they were established. Because universities legitimize extractive and exploitative systems, they contribute to multiple types of injustices, including climate injustices and ecological devastation. One key mechanism for reinforcing injustice is the intentional exclusion of certain kinds of people and certain kinds of ideas. Institutional success and university reputations are often defined by how competitive and exclusive they are. Another kind of injustice is the restriction of certain kinds of inquiry and the prioritization of specific types of knowledge over others. Cultural practices that rely on explicit and implicit intimidation are widespread, including everything from aggressive and hostile questions during university seminars (Dupas et al. 2021) to hiring and admissions decisions that are structured to disadvantage social justice advocates.

Despite the promise of academic freedom, many academic staff do not feel free to speak up on their campuses. A 2023 report in the United States found that a third of faculty self-censor out of concern over responses of staff, students, or administrators, and 91% are at least "somewhat likely" to self-censor in meetings, presentations, publications, and/or on social media (Honeycutt, Stevens, and Kaufmann 2023). The growing economic precarity of many people working in academia, coupled with the competitive, stressful, gendered, and racialized environment, diminishes the potential for individuals to challenge dominant systems. The erosion of faculty governance (Schoorman 2018) and the corporatization of university management (Washburn 2005) further discourage and disempower institutional change in response to students, faculty, and staff within. Unfortunately, many academic institutions have evolved to reinforce their relevance not by rapidly adapting to the ever-changing needs of society but instead by upholding a rigid hierarchy of knowledge and intensifying individualistic and institutional competition.

This chapter synthesizes a broad range of critiques of higher education, focusing on: (1) patriarchy, misogyny, and gender-based violence; (2) structural racism; (3) coloniality; (4) capitalism, corporatization, and financialization; (5) climate injustices and climate coloniality; and (6) the undermining of well-being. This review draws from, and contributes to, the growing field of critical university studies, which questions how universities are contributing to society and explores the role of higher education institutions in upholding and reinforcing hegemonic paradigms (Boggs and Mitchell 2018). From their earliest existence, universities and formal education systems have been criticized by scholars for their role in society (Wagner, Acker, and Mayuzumi 2008; Newman 1893; Eells 1934).

Many changes within higher education, including the establishment of new universities, have emerged in response to criticisms of existing higher education practices. The first Catholic University of Ireland was founded in 1854 by John Henry Newman, an English theologian, poet, and cardinal who advocated for a comprehensive, liberal, interdisciplinary university, a place for teaching universal knowledge. His ideas, published in the 1893 publication *The Idea of a University*, reflected his criticism of the trend toward specialization and utilitarianism that he saw at Oxford University; his mission was to provide an institution that developed holistic intellectuals with critical thinking skills (Newman 1893; Ker 2011). A more recent example is the establishment of the College of the Atlantic in Bar Harbor, Maine, in the northeastern United States, which was founded in 1969 by peace activists who envisioned a new model of higher education that combines practical experience with an academic focus on complex interactions between humans and their natural, cultural, and built environments.

The field of critical university studies has expanded to understand the impacts of the increasing financialization of higher education, that is, the growing influence of financial motives and practices in shaping universities' priorities, governance structures, and funding streams (Eaton 2022). Scholars are exploring how current structures and incentives within academia are reinforcing problematic power dynamics and constraining how higher education institutions can engage with humanity's greatest challenges (Kelly et al. 2022; Quadlin and Powell 2022; Russel, Smith, and Sloan 2016). Hierarchal, financialized institutions that uphold power differentials among people by leveraging fear of economic precarity constrain creativity in teaching, learning, and research. Even the common phrase *higher education*, which is often used to describe universities, community colleges, technical institutes, and other institutes of third-level education (a phrase that I have chosen to use extensively in this book to refer to this array of different kinds of educational institutions), demonstrates the explicit and fundamental hierarchal assumptions within education.

Many of those whose research and writings are critical of higher education institutions recognize the loss to society resulting from the exclusion associated with academic structures and systems (Davies 1866; Fitzpatrick 2019; Byrd 2021). In solidarity with so many of these other critical scholars and activists, I feel a deep sense of responsibility and possibility regarding a very different kind of impact that academic institutions could have as climate destabilization worsens vulnerabilities in our interconnected world. Here I review multiple critical perspectives to set the stage for the reimagined alternatives in the subsequent chapters.

Patriarchy, Misogyny, and Gender Violence in Higher Education

Patriarchal sexist structures are so pervasive and mainstream in universities around the world that many students, faculty, and administrators simply accept and adapt to them. Academic institutions reinforce gender oppression in multiple ways, including gender-based discrimination, sexual harassment, and sexual assault (Wagner and Yee 2011). The gendered terrain of higher education is pervasive and includes gender bias in admissions, limited representation in leadership, gendered curriculum and pedagogies, and unequal access to resources, opportunities, and career progression (Parkes 2004; Wagner, Acker, and Mayuzumi 2008). Academic institutions have also played a major role in reinforcing societal tendencies of genderism, the rigid adherence to a gender binary in practices, policies, discourse, and norms (Marine and Nicolazzo 2014).

The many challenges transgender students, faculty, and staff face in higher education institutions—including accessing inclusive health care; experiencing discrimination, prejudice, and violence from peers; and dealing with a lack of inclusive policies, support, and resources—results in exclusion and isolation that can further disadvantage them professionally, educationally, and financially (Lennon and Mistler 2010; Annie 2017; Marine and Nicolazzo 2014). Cisnormativity, the perpetuation of the narrow belief that there are only two genders, that gender cannot be changed, and that bodies define gender (Simmons and White 2014), is deeply embedded in educational systems around the world (McBride and Neary 2021).

Unfortunately, despite multiple laws, regulations, and trainings designed to reduce sexual harassment in academia, many universities have resisted changes, and most address only the minimum required by law (Hall 2021; Tenbrunsel, Rees, and Diekmann 2019). Because of this, there remains a staggering number and steady drumbeat of publicly disclosed cases of sexual harassment and gender-based violence in academic institutions around the world (Pritchard and Edwards 2023; National Academies 2018; Towl and Walker 2019). Comprehensive data characterizing the prevalence of sexual misconduct in higher education does not exist because many incidents—including my own sexual harassment experience and claim that I mentioned at the beginning of this chapter—are dealt with internally with no external acknowledgement or reporting. Nevertheless, a 2020 review concluded that sexual harassment is a global epidemic throughout global higher education systems (Bondestam and Lundqvist 2020). This study identified a major gap in theoretical, longitudinal, qualitative, and intersectional research on the prevalence and impacts of these abuses of power (Bondestam and Lundqvist 2020). For example, a 2023 study of academic medicine in Germany found that 70% of medical providers experienced sexual misconduct during their university training, but the study did not offer any insights into how the misconduct impacted the individuals or the organizations where it occurred (Jenner et al. 2019). In the US context, research has shown a higher occurrence of sexual harassment in academia than in the private sector or government—the military is the only sector with more incidents than US academia (Ilies et al. 2003).

The professional and personal impacts of widespread sexual harassment in higher education are not generally recognized or well understood in most universities; this lack of awareness results from the patriarchal structures that privilege the established individuals in the hierarchal system (Pritchard and Edwards 2023). Minimizing the impact and experiences of victims allows sexual extractivism to persist. A claim frequently made by victims of sexual violence on campus is that universities take plagiarism more seriously than they do rape (Li 2014).

Sexual coercion within the university hierarchy is a persistent form of academic gatekeeping and intimidation. The practice of senior academics rewarding subordinates who respond favorably to pressure to engage sexually and penalizing those who do not was described in a powerful exposé in a 2023 publication coauthored by three academic women who attended the same university and were subjected to the same sexual extractivism (Viaene, Laranjeiro, and Tom 2023). In addition to sexual harassment of subordinates, research in the 1980s revealed that women professors experienced a range of sexual harassment behaviors, not only from their colleagues, but also from their students (Grauerholz 1989).

When high-profile stories of academic sexual harassment make it to the public sphere, the reporting often still focuses on the decline of the academic star—and not on the impact on those who have been abused (Zelikova, Ramirez, and Lipps 2018). Kate Mann describes the widespread phenomenon of "himpathy"—that is, expressing more empathy for the man in a conflict—which is a prevalent, reinforcing part of patriarchy frequently expressed in academic institutions (Manne 2018). The hierarchies of higher education also create environments for sexual harassment and abuse of men by women, of people by the same gender, and of trans people.

Given how effective academic institutions are in protecting their reputations and reinforcing patriarchal norms, investigative journalism has played a critically important role in exposing the pervasive and structural problems of gender-based violence and abuse in higher education. Multiple media outlets, including the *Guardian* (Hall 2021), Al Jazeera's *Degrees of Abuse* series (Bull 2021), and extensive reporting for years by Nell Gluckman and others at the *Chronicle of Higher Education* in the United States (Gluckman 2017), have kept this issue in the public eye despite intensive efforts by academic institutions to ignore, dismiss, and diminish the devastating impacts of abuses of power within their sector. Although the #MeToo movement—and its counterparts throughout the world, including #YoTambien in Latin America and Spain and #KuToo in Japan—opened up space for more intentional sharing of stories about sexual predators in the workplace, the rigidity of academic structures has meant that abuses of power and silencing within higher education remain pervasive (Bull 2021).

Acceptance of gender-based discrimination, racialized exclusion, bullying, intimidation, and abusive relationships within academic institutions is often assumed to be part of the rigor associated with the intellectual expertise that professors offer their students and subordinates (Kim and Xu 2023). Many contemporary academic institutions retain the patriarchal structures associated with the controlling, violent cultures of early universities in medieval Europe (Rudy 1984). Early universities were formed as academic guilds that provided protection and a hierarchy for learned men who taught other men theology, law, and medicine (Rudy 1984). In these institutions, and in many universities today, intellectual arrogance and a sense of superiority is nurtured. This arrogance reinforces a competitive environment that rewards those who are narrowly focused on demonstrating their individual academic success.

University faculty in many parts of the world are still dominated by white men; white male leaders make up most of the senior administrative positions in North American, European, and Australian universities (Fleck 2022). Because of the legacy

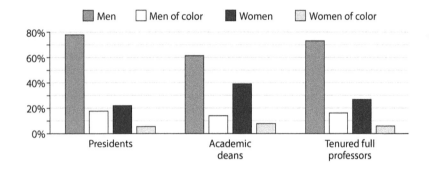

Figure 3 White Male Leadership Still Dominates US Higher Education. Percent of men and women in leadership positions, based on America's top 130 research universities, ranked by the Carnegie Classification as of September 2021. *Source:* Women's Power Gap Initiative, https://cdn.statcdn.com/Infographic/images/normal/27336.jpeg.

of exclusionary structures, including the long tradition of excluding women (Carlton 2023), universities perpetuate a narrow focus on certain disciplines of knowledge and reinforce traditional patriarchal cultures. A systemic gender bias is widely recognized in academic institutions around the world: women hold only 22% of leadership positions in US research universities (Fleck 2022) (figure 3), and in Europe only 24% of institutional leadership roles are held by women (Galligan 2022). Only 24% of academic staff across sub-Saharan Africa are female, and in Ghana women make up only 8% of professors at public universities (ESSA 2021). While women make up half of all academics in Chinese higher education institutions (Tang and Horta 2021), only 4.5% of China's higher educational institutional leaders are women (Zhao and Jones 2017). The data demonstrate that despite the many laws, policies, initiatives, and trainings designed to diversify academic leadership, prevent sexual harassment, and promote gender equity and inclusion, academic structures continue to reinforce and enable gender-based violence and discrimination.

If pledges and goals for gender equity are ever to be realized, deeper structural changes in how universities are organized and managed are required. Without systemic change in the traditional hierarchies and patriarchal cultures and practices in universities, continued occurrences of sexual violence, gender-based discrimination, and toxic environments in higher education should be expected.

Structural Racism in Academia

Racial exclusion, discrimination, and violence in higher education is just as pervasive and arguably more persistent than gender-based abuse and sexual harassment. In different cultures and contexts around the world, structural racism shows up differently, so the ways that academia reinforces structural racism varies.

In the United States, institutional racism within higher education is widespread and the legacy of racial segregation persists (Bracey 2017). Historically Black colleges and universities (HBCUs), institutions established to educate Black Americans who were banned from white colleges and universities before the Civil Rights Act of 1964, remain underfunded and marginalized compared to predominantly white institutions (PWIs). According to a 2022 White House "Proclamation on National Black Colleges and Universities Week," HBCUs have educated 40% of Black engineers, 50% of Black lawyers, 70% of Black doctors, and 80% of Black judges. Recognizing the stark racial inequities in health, wealth, and educational outcomes in the United States, many predominantly white universities have been expanding their racial justice initiatives and their investments in diversity, equity, and inclusion (DEI) programs and trainings. These efforts intensified with the rise of the Black Lives Matter movement and following widespread protests after the brutal murder by police of George Floyd in

2020. As universities try to promote racial equity on their campuses, the ineffectiveness of these institutional efforts to change the structural racism embedded within academia is becoming increasingly clear (Byrd 2021; MacKenzie et al. 2023). Investing in performative DEI programs and making minor policy changes without changing institutional values and structures have been insufficient and inadequate (Abrica, Hatch-Tocaimaza, and Rios-Aguilar 2021).

In response to the violent acts of racism that caused the deaths of countless Black Americans in 2020, Northeastern University's leadership joined the voices of many university presidents in the United States to announce a renewed commitment to DEI as part of a call to action for the university "to address the scourge of systemic racism." Part of this renewed commitment included establishing a new goal of increasing the diversity of students, faculty, and staff at the university to reflect the diversity of the population of the United States. Without systemic or structural changes, however, it is not clear whether, how, and when goals like this will be met.

Growing critiques of DEI efforts in higher education have pointed out that many diversity initiatives focus the burden on individuals to make changes to fit the existing institutional structures rather than on changing the organizational structures to become more welcoming and inclusive to diverse individuals (Byrd 2021). This individualistic rather than systemic approach to addressing structural racism results in "identity taxation" in academia, which means additional labor is expected to be performed by faculty from marginalized groups, adding yet another burden that can detract from time spent on their other responsibilities (Joseph and Hirshfield 2023).

The systemic exclusion of non-white and marginalized scholars and students in many university systems perpetuates white supremacy, the ideology or belief system rooted in Europe and its colonial history that assumes white people, including white culture, values, and achievements, are superior to other racial or ethnic groups (Yacovone 2022). A lack of diversity in higher education also discourages future diversity because young people can be discouraged when they do not see underrepresented scholars serving as role models inspiring them to follow (Tillman 2018). Systemic exclusion also perpetuates racial economic disparities because individuals from certain racial, ethnic, and cultural backgrounds face structural barriers to accessing higher education. This exclusion also perpetuates socioeconomic inequities, with long-term impacts on the economic and social well-being of marginalized communities. Exclusion of any group or community from higher education deprives society of the diverse perspectives and talents that could be contributed through academic activities. Diversity of all kinds enriches innovative thinking and creative learning (Stewart and Valian 2018; Taylor 2018), so systemic exclusion of people based on race, gender, or other cultural identities diminishes the university's potential to contribute innovative approaches to research, teaching, and engaging with the world. Excluding some groups of people while privileging others constrains academic contributions and the scope of research. Racial exclusion in environmental research delayed the identification of environmental racism (Bullard 1993) and continues to minimize attention to environmental injustice. Widespread concern about the disproportionate impact of environmental damage on non-white communities has expanded because of the contributions of non-white scientists who have prioritized documenting the health disparities and inequities in ecological harm (Bullard and Johnson 2000; Stephens 2020).

Universities in the United States, Canada, and many parts of Europe have become more accessible to more people over time; however, people from wealthy, white, European backgrounds who have traditionally perpetuated their privilege through higher education institutions continue to have greater access in most contexts. The title of a 2004 article by US psychology scholars Edward Renner and Thom Moore, "The More Things Change, the More They Stay the Same: The Elusive Search for Racial Equity in Higher Education," reflects how efforts to expand racial equity in the US educational system have created an illusion of change while reinforcing white advantage and privilege. Despite all the efforts universities have made in the more than twenty years since that article was written, the same illusion of change persists. Numbers reveal continued white privilege (Byrd 2021) despite specific initiatives, including changes to university admissions processes (Moses and Jenkins 2014; Bleemer 2023), mentoring of faculty of color (Tillman 2018), and expanding support and resources for non-white students, faculty, and staff (Nunes 2021). Although universities often portray themselves as equity conscious by using language of anti-oppression and anti-racism in their policy documents, websites, and recruitment materials, this rhetoric is often not supported with institutional actions and changes to organizational structures and processes (Wagner and Yee 2011). Growing calls for deeper restructuring and open acknowledgement of the cultural and intellectual gatekeeping that persist in academia are aligned with the larger transformative paradigm shift proposed with the concept of climate justice universities.

An initial critical step to disrupt the many ways that higher education perpetuates white supremacy, patriarchy, and a mindset of exploitation requires universities to be self-reflective and transparent while holding themselves accountable. Rather than covering up and avoiding public discourse about their role in structural racism, sexism, and climate injustices, universities could interrogate their own histories, collectively question their own values, and transparently interrogate their influence on society.

One prominent example of this is Harvard University's interrogation of its institutional ties to slavery. In 2022, a committee of Harvard faculty convened by the university president released a thoroughly investigated report revealing the many ways that the institution had exploited enslaved people and benefited financially from slavery (Harvard 2022). Among the many findings included in the report was the acknowledgement that slavery was part of daily life at the university for over 150 years. "Enslaved men and women served Harvard presidents and professors and fed and cared for Harvard students" (Harvard 2022). The report also revealed that the university was dependent on wealth from slave trading, plantation owners, and direct financial investments in slave-dependent production of cotton, sugar, and rum. A strong resistance to calls for racial integration persisted as Harvard leaders tried to sustain a university that only served white, wealthy men. Harvard faculty contributed to this resistance by disseminating bogus scientific claims of racial difference that were used to justify exclusion of Black men. Louis Agassiz, one of the most well-known Harvard professors in the 1850s and 1860s, who popularized science by establishing a Harvard Museum, used his science on the origins of species to argue that non-white people were part of a different race that was inferior to white people (as previously mentioned in the introduction). Agassiz was a prominent voice advocating for segregation and against mixed-race children, and his views on race were influential in the formalization of segregation policies in the United States.

As part of Harvard's Legacy of Slavery Initiative, the university recognized the institutional need to take reparative action. With the release of the 2022 report, the university pledged $100 million to create an endowed fund to "redress" past wrongs.

A committee was formed to allocate these funds to support teaching, research, and service that might "redress" Harvard's legacy (Bacow 2022). Dania Francis, a University of Massachusetts economist who studies racial economic disparities and the economics of education and stratification, has pointed out that $100 million is "a drop in the bucket relative to the $11 trillion racial wealth gap" in the United States (Leung 2022). This number is also small compared both to the hundreds of millions that Harvard received in philanthropic donations in 2023 and to Harvard's endowment, which was reported to be $50.7 billion in 2023. While Francis recognizes that local and private initiatives designed to address the legacy of slavery are well-meaning, she highlights that these individual institutional efforts detract from national reparation initiatives. Rather than focusing inward on itself, some argue that Harvard's time, effort, and money could be better spent supporting and contributing to larger efforts to promote reparative actions throughout the United States.

The small scale of this reparations initiative at Harvard demonstrates the insufficiency of individual universities acting in isolation focused narrowly on themselves. Rather than engaging more directly with systemic and structural change in society or contributing to collective national or global efforts to advance reparative justice, Harvard chose to make a small investment in itself. While in some respects Harvard's efforts to reveal and redress its legacy of slavery demonstrate institutional courage and racial justice leadership, its institutional actions are nontransformative, largely symbolic, and self-serving.

Coloniality in Higher Education

Another area of critique of higher education focuses on coloniality. Many academic institutions were established by colonial powers to serve white men and advance colonial interests

(Stein 2022). Colonial powers justified their actions by dehumanizing others and elevating their own priorities and their own welfare as more important than the priorities and welfare of the people who lived in their colonized states before they arrived. Coloniality describes how the oppressive and exploitative power dynamics of colonialism continue even after colonialism has ceased (de Onís 2018). Coloniality is perpetuated through the educational systems established during colonialism because the legacy exclusionary practices and the dehumanizing priorities remain; acknowledging this is a critical first step to promoting equity and justice (Nayak 2023).

Settler colonialism refers to the specific form of colonization in which colonizers established permanent settlements on lands where indigenous people lived, violently displacing them. Many universities, particularly in settler-colonial places, including the United States, Canada, Australia, Israel, and New Zealand, were strategically established as part of the land appropriation and colonizing process to consolidate control of the land and resources. Universities also marginalized indigenous knowledge and traditions by promoting European knowledge systems, languages, and cultural norms. These universities implemented policies that explicitly excluded indigenous people, thereby reinforcing the dominance of the settler societies and the colonial power structures.

The curriculum and knowledge systems of universities around the world are still Eurocentric—that is, European perspectives and knowledge are prioritized over those of other cultures and civilizations. By presenting Western or European knowledge systems as superior to indigenous or other local knowledge systems, universities have devalued and undermined local knowledge systems and contributed to the erasure of indigenous cultures and their ways of knowing. Many settler-colonial universities have culturally appropriated indigenous symbols, names, and mascots to represent the university; this practice perpetuates stereotypes about indigenous people and contributes to further marginalization.

As the coloniality of educational systems has become more widely recognized (Mbembe 2021), efforts to decolonize schools and universities (Cortina et al. 2019; Alvares and Faruqi 2012; Bhambra, Gebrial, and Nisancioglu 2018), to decolonize student thinking (Adefarakan 2018), to decolonize the curriculum and pedagogy (Shahjahan et al. 2022), and to decolonize how academia theorizes, teaches, and researches community (Dutta 2018) have expanded around the world in many different contexts. The 2018 book *Decolonising the University* explores theory and practice of decolonization, recognizing that Western universities continue to be key sites where colonialism—and colonial knowledge in particular—is produced, consecrated, and institutionalized (Bhambra, Gebrial, and Nisancioglu). As such, the unsettling and displacing of Eurocentrism in universities is a key part of decolonization processes (Cupples and Grosfoguel 2019).

In response to the pervasive use of the term *decolonization* in education, critical scholars Eve Tuck and K. Wayne Yang, in their influential 2012 article entitled "Decolonization Is Not a Metaphor," pointed out that decolonization is a material struggle over stolen land and that educators should engage more with the actual physical processes related to land rather than the symbolic. When the term is appropriated and diluted in education, activism, and research, Tuck and Yang argue, the ongoing harmful realities of colonialism and coloniality can be obscured. This perspective highlights the importance of approaching decolonization by centering the lived experiences and struggles of people and communities marginalized by colonialism. Ubiquitous and shallow application of this word can further entrench coloniality by minimizing the struggle and suffering.

Parallel arguments have been made about the ubiquitous efforts to make claims of advancing racial justice but then implementing nontransformative DEI programs that do not address systemic and structural inequities (Porter, Wang, and Dunn 2023). Many DEI programs dilute, distract, and minimize racial justice struggles and thus further entrench racial inequities (Joseph and Hirshfield 2023; Byrd 2021). When specific language and ambitious rhetoric is used but the actions taken are minor, performative, and nontransformative, the minor incremental actions can serve to reinforce the problematic power structures by diminishing the impression of the oppressor's continued oppression (MacKenzie et al. 2023). Investing in high-profile public efforts to support a more diverse and inclusive organization without actually changing some of the fundamental exclusionary structures is a form of distraction or obstruction. Similar to the greenwashing climate delay tactics of fossil fuel companies that prioritize renewable energy transitions in their public messaging while simultaneously demonstrating no intention of phasing out their exploration and extraction of fossil fuels (Si et al. 2023), many higher education institutions are making bold claims of advancing racial justice and investing in racial equity without any plans to change the exclusive systems that they perpetuate.

Capitalistic Corporatization and Financialization: The Academic-Industrial Complex

Universities around the world are increasingly influenced by capitalism, corporatization, and financialization. Capitalism, a term increasingly associated with the destructive path humanity is currently on, refers to a specific economic and social system based on private ownership of resources, production dominated by profit maximization of individuals and businesses through free market exchanges (Oksala 2023). This unique system of market dependence relies on competition, accumulation, profit maximization, and ever-increasing labor productivity, which shapes not only economic interactions but also social relations and the relationships that people have with the nonhuman world (Meiksins Wood 2017). One fundamental problem of capitalism is that it privatizes gains and profits (economic gains are allocated to specific individuals and organizations) but it socializes the damages and losses (harm is distributed collectively throughout society) (Gosha 2022). Expanding research shows that capitalism, and the capitalist growth imperative, is the root cause of the climate crisis (J. Green 2022; Hickel 2021; Hickel et al. 2022; Klein 2011; Klein 2014; Newell and Paterson 2010; Oksala 2023; Ramanujam 2023; Speth 2008). This direct link between capitalism and climate injustice means that climate justice universities must engage directly with their own relationship with capitalism and strive toward a paradigm shift in the corporatization and financialization of academia.

The corporatization of higher education refers to the trend of universities being increasingly managed like businesses (Engelen, Fernandez, and Hendrikse 2014; Washburn 2005). By increasingly deploying profit-seeking entrepreneurial practices, universities act more like economic enterprises, constantly trying to maximize their revenues and advance their own academic competitiveness rather than maximizing the societal impact of their academic activities (Jessop 2018).

Financialization is a general term to describe the ever-expanding role of finance and debt in society, the focus on accumulating money above all else, and the process by which financial markets, financial elites, and financial institutions gain greater influence over policies and practices (Lapavitsas 2013; Mader, Mertens, and van der Zwan 2020). The financialization of higher education refers to the increasing influence of financial motives and practices in shaping the organization, governance, and funding of universities (Eaton et al. 2016). The financialization of higher education has been accelerated by insecure and, in many places, decreasing public funding for higher education.

The decline in public funding for higher education is a response to the financialization of the public sector and government entities. With reduced public funding, many higher education institutions have had to find other sources of financial support. This need has opened doors for corporate influence and wealthy donors, including the fossil fuel industry, to influence university priorities and research agendas. This growing influence of corporate and wealthy donors has strengthened opportunities for higher education institutions to be leveraged for the accumulation of wealth and power. This is part of a reinforcing cycle because the same large corporations and wealthy donors who contribute to universities are in many places also strategically advocating for smaller government, lower taxes, and less public funding for education. The strategic undermining of public services and weakening of the government that historians Naomi Oreskes and Erik Conway so effectively describe in their 2023 book, *The Big Myth: How American Business Taught Us to Loathe Government and Love the Free Market*, is linked to the financialization of higher education and the lack of independent thinking within universities.

Financialization of higher education has grown so extreme in the United States that one critic of Princeton University characterized the elite private university as "a hedge-fund that conducts classes" (Young 2016). Others have called out universities as "real-estate developers" (Bula 2017) or "giant piggy banks for hedge-fund billionaires" (Liu 2023). The constant prioritization of financial returns and how best to compete for students and external research funds constrains the kinds of educational programs that are offered, the type of research that is supported, and who has access to higher education. When institutions prioritize their financial returns and compete within the higher education market to attract students by investing in attractive buildings and physical infrastructure, investments in academic programs and instructional staff are often viewed as less important. This not only diminishes time and space for academic creativity, but it also delegitimizes academic work when academic freedom is being so blatantly leveraged for financial gain. Research shows that the financialization of higher education erodes the extent to which anti-oppression teaching can be expected to challenge the existing order (Wagner and Yee 2011).

The term *academic capitalism* refers to the increasing commodification of knowledge production and consumption and the marketization of education (Slaughter and Leslie 1999). Many universities are increasingly treating students as consumers from whom profit can be extracted and treating academic staff as education workers from whom ever more productivity can be demanded (Barry 2011). Not only is the rise of academic capitalism demonstrated in the competitive landscape of universities trying to recruit students, it is also evident in the high profits from the for-profit publishing companies that own the academic journals where academic research is published (Connell 2019).

To maintain their competitive edge in the higher education marketplace, universities are increasingly seeking financial support from wealthy individuals and organizations, often catering to corporate interests in terms of both knowledge dissemination and knowledge creation. The expanding close relationship between academia and industry, including their collaborations in research, development, and innovation, has been characterized as a growing *academic-industrial complex* (paperson 2017). This concept—an adaptation from the more well-known phrase *the military-industrial complex*, which describes the interdependence between the defense industry and militaries around the world—reflects criticism and concern that funding university research that benefits corporations is taking precedence over funding valuable research areas that lack immediate economic returns. Each week there seems to be additional empirical evidence demonstrating just how insipid and central corporate and elite capture of higher education has become (Kumar 2023). An example of this is the strategic investments that the fossil fuel industry has been making for decades to fund university-based climate and energy research (discussed more in chapter 4); a growing body of research shows how this industry-sponsored academic funding is part of the fossil fuel industry's climate obstruction efforts to deny, delay, and distract from climate policy that would force urgently needed fossil fuel phaseout (Supran, Rahmstorf, and Oreskes 2023).

The concept of the academic-industrial complex also reflects the growth of predatory for-profit colleges (Cottom 2017) and the financial gains of institutions that partner with industry (paperson 2017). The concept also helps explain why the fossil fuel industry and other corporate interests have contributed to the erosion of public support for higher education; they have created the financial dependence that gives them power and influence in higher education institutions. Limited and ever-reduced public funding of universities in many parts of the world is requiring universities to increasingly look to the private sector or wealthy donors for financial support. This increased reliance on philanthropic and corporate financing has changed the incentive structure within higher education with regard to teaching, learning, research, and engagement. The constant pressure to bring external financial support to higher education institutions has resulted in disproportionate interest in technological innovation and private entrepreneurship and diminished research interest in social innovations that serve the public good (Boyle and Stephens 2022).

Because capitalism relies on patriarchy and white supremacy and other "interlocking systems of oppression"—a concept introduced to social movements by the Combahee River Collective in 1977 (Carastathis 2016)—links between capitalism and racism are increasingly recognized. The concept of racial capitalism describes the mutually supportive relationship between racism and capitalism (Táíwò 2023). With some universities now declaring a commitment to become antiracist organizations, the case could be made that those universities should also commit to becoming anti-capitalist organizations. To my knowledge, there are no universities that claim to be anti-capitalist, although Sterling College in Vermont embraces a degrowth model of education. Reclaiming and restructuring the financial architectures of higher education is an essential part of reimagining climate justice universities; this is the focus of chapter 5.

Climate Injustice and Climate Coloniality in Academia

Yet another area of critique of academia relates to climate injustice and climate coloniality. We know that more frequent and extreme disruptive climatic events are adversely impacting water access, food production, physical and mental health, and physical and economic infrastructure—particularly for vulnerable communities around the world (IPCC 2022b). Yet most higher education efforts to address the climate crisis are narrowly focused on incremental and nontransformative changes; increasingly performative efforts at universities are being called out as greenwashing (de Freitas Netto et al. 2020). In this new era of multiple intersecting, globally connected injustices, higher education's climate commitments need to link to social justice, health equity, and economic justice (Harlan, Pellow, and Roberts 2015; Chankseliani and McCowan 2021). Because climate devastation is experienced differently among vulnerable people, marginalized communities, and peripheral places, universities are perpetuating injustices when they engage with the climate crisis without making the conceptual and empirical links to social injustices, economic inequities, and health disparities (Cappelli, Costantini, and Consoli 2021; Singer 2018).

Countries and communities that have contributed the least to the climate crisis are among the most vulnerable (Watts et al. 2021). Extreme heat, storms, flooding, drought, and unpredictable weather of all kinds are disrupting livelihoods and food production. Climate disruptions, caused disproportionately by the excessive consumption of wealthy individuals, sometimes referred to as "the polluter elite" (Kenner 2019), are exacerbating armed conflict and forcing climate migration (United Nations 2019). Many of these inequitable vulnerabilities result directly from colonial legacies that violently co-opted environments and resources for unsustainable extraction (Howitt 2020; Sultana 2024). Higher education institutions are powerful organizations reinforcing the interlocking systems of oppression that are worsening climate injustices.

The role of universities in exacerbating climate injustice is at least in part related to higher education's endorsement of climate isolationism, a term I coined to describe the common framing of climate change as an isolated, discrete, scientific problem in need of economic and technological solutions

(Stephens 2022). The climate scientists and engineers within the higher education sector have been instrumental in legitimizing this narrow technocratic, nontransformative approach to climate policy that has proven to be inadequate and ineffective. Focusing almost exclusively on quantitative greenhouse gas emission reductions and temperature change, while inadvertently ignoring the societal complexities and potential social innovations, university-endorsed climate isolationism has diminished the potential for transformative social change (Anderson and Peters 2016). When the climate crisis is framed as a scientific problem in need of a technological fix, public discourse on and imagination for changing the underlying societal and economic structures are severely constrained. The role of the university in promoting and legitimizing climate isolationism is a form of climate obstruction (Ekberg et al. 2022; Lamb et al. 2020) because it delays transformative action and distracts from the visioning of alternative social and economic futures (Stephens 2020).

The silos of higher education have perpetuated climate isolationism by emphasizing and supporting physical science and technological innovation to address climate change. Despite efforts to diversify science and engineering, persistent racial, gendered, and economic injustices of our economy and educational systems perpetuate exclusive educational access to fields of science and engineering (Valantine and Collins 2015). The lack of diversity within these fields limits the scope of inquiry and constrains the types of connections that are made among science, technology, and society (Stephens 2020). Many colleges and universities sustain patriarchal leadership structures as they promote technocratic individualistic goals that prioritize the future financial success of their students, alumni, and partnerships that serve corporate interests. The financialization of higher education has limited institutional commitments to prioritizing the public good and civic engagement.

The problematic influence of the private sector and corporate interests in higher education is clear when one considers how and why fossil fuel companies have strategically supported higher education research since the 1950s (Westervelt 2021). The influence of the Koch family on higher education research (Leonard 2019) is the most widely recognized higher education funding source promoting climate denial and strategically resisting climate action. A larger network supporting the climate change counter movement (CCCM) has focused on providing financial support to colleges and universities in order to strategically resist climate action and undermine efforts to reduce fossil fuel reliance (McKie 2021; Westervelt 2021). Because higher education has been prioritizing private interests over the public good—the collective good of society—by accepting funds to amplify climate denial and promote climate isolationism, what is needed are transformational changes in the functioning of higher education and how it interacts with corporate interests, the public sector, and marginalized communities.

While technology is an essential part of a transition toward a more just, equitable, and climate-stable future, investments in physical science and technological innovation have not yet been adequately balanced with investments in social science, social infrastructure, social innovations, and social justice (Overland and Sovacool 2020). This lack of investment has weakened social ties, thereby reducing community resilience—our ability to collectively cope with, and recover from, crises (Aldrich 2012). The lack of investment in social innovation and social justice has also constrained our imaginations when it comes to the role and potential impact of higher education in society. This narrow approach has begun to shift as there is growing recognition that addressing climate change will require investing in transformative social, institutional, financial, and political changes (Overland and Sovacool 2020). Still, data on research grants and funding shows that higher education continues to emphasize climate research in the fields of science and technology rather than climate research in the social sciences such as economics, politics, and sociology (Overland and Sovacool 2020). If social science research and social innovation were prioritized and funded at a higher level, technological innovation would be coupled more effectively with research to accelerate the accompanying social change. As influential innovative institutions, higher education has an opportunity to lead by example and change the discourse from a climate isolationist approach to a more holistic and integrated commitment to climate justice across campus functions and initiatives.

For higher education to move beyond narrow climate isolationism and instead move toward climate justice universities, it needs to commit to addressing the underlying injustices and inequities that contribute to climate vulnerabilities. This requires recognition that colleges and universities shape the communities in which they are located. Providing good jobs and economic vitality is often an assumed role of higher education institutions in their local communities; however, "town-gown" interactions and relationships are often contentious, particularly when the college or university seems to be extracting from, rather than contributing to, the local community (Mtawa and Wangenge-Ouma 2022; Baldwin 2021).

As community-engaged research and experiential learning are increasingly encouraged in higher education, colleges and universities are reckoning directly with the results of exploitative and extractive relationships with local communities (Riccio, Mecagni, and Berkey 2022). In the United States, the public good of higher education institutions is legally and fiscally recognized through their tax-exempt status (universities do not have to pay property tax to the city or town where they are located). This tax exemption has become increasingly controversial because local municipalities are disadvantaged by their lower tax base when higher education institutions expand and accumulate more and more land, reducing the amount of local property tax received by their host communities (Baldwin 2021). Within this current framework, many universities in the United States end up contributing more economic benefits to the private sector than they do to local communities (Quigley 2018; Baldwin 2021).

A PILOT (payment in lieu of taxes) assessment program, in which universities voluntarily contribute to local municipalities, has been established in some places in the US to compensate for the loss in tax revenue. PILOT is designed as a direct economic contribution from the university to the local community; however, many universities decide to contribute less than the amount recommended in the PILOT assessment, and because the program is voluntary, the host communities have no recourse when universities minimize their contributions (Quigley 2018). Expanding direct investments by higher education in funding public infrastructure used by both the campus and surrounding community could simultaneously advance multiple goals, including those of climate justice. Examples include higher education investing in fare-free public transit, upgrading local water infrastructure, building efficiency, installing community-based clean energy microgrids, and contributing to community resilience initiatives in anticipation of more frequent and intense climate disruptions.

Climate coloniality describes how the hierarchical imbalances of power created by coloniality are perpetuating the climate crisis (Sultana 2022b; Sultana 2024). Colonial models of extraction of materials and exploitation of labor strategically designed to concentrate wealth and power among colonizers are reflected, reinforced, and reproduced in current neoliberal extractivist imperial structures and institutions, including those of higher education institutions. The climate crisis is replicating the patterns of colonialism that rely on dehumanizing others and prioritizing the well-being and primacy of the colonizer over the colonized. Normalizing the idea of expanded human suffering and species extinction is part of climate colonialism. Many proposed approaches and current strategies for taking action on the climate crisis are nontransformative, elevating the priorities of those most privileged while minimizing the priorities of those most vulnerable. Acknowledging universities as central nodes of climate coloniality provides a compelling framework for justifying the need to resist, reclaim, and restructure the current higher education system to allow for decolonial, anti-racist, feminist practices and priorities to thrive in order to further the goals of climate justice.

The fossil fuel industry's significant involvement and influence in higher education around the world (Franta and Supran 2017) demonstrates how universities are contributing to climate injustice. Fossil fuel companies (and their affiliated foundations) fund climate and energy research, host student recruitment events at campuses, sit on university governance boards, and leverage their funding to influence program development and curriculum decisions. Public discourse on the links between the fossil fuel industry and higher education has historically focused on the fossil fuel divestment movement (Healy and Debski 2017). Since at least 2020, however, these campaigns have widened their focus to "kicking oil companies out of school," calling on their institutions to reject fossil fuel funding and other partnerships (Tabuchi 2022). Journalists, civil society organizations, and student groups have provided detailed accounts of these partnerships in an attempt to highlight the dangers of corporate influence on university teaching, research, and administration. However, systematic scholarly research on the scale, consequences, and resistance to this phenomenon is limited (Hiltner et al. 2024). The report released in 2023 by the organization Data for Progress identified the significant influence of fossil fuel money in academia. Eradicating this influence needs to be an integral part of building climate justice universities.

The powerful influence of fossil fuel interests in academia is related to the patriarchal and misogynist culture associated with both universities and fossil fuels. The term *petromasculinity*, coined by feminist political scientist Cara Daggett, describes how fossil fuel systems are associated with white, patriarchal authority (Dagget 2018). This concept suggests that clinging to fossil fuel futures is not just based on profits but it is also often linked to identity. For higher education institutions with strong white supremacist, patriarchal legacies and structures, therefore, partnering with fossil fuel industry actors may offer some university leaders more than just financial support.

Undermining Well-Being in University Systems

A final critique of contemporary academia relates to the well-being of employees and students. Universities becoming increasingly corporatized with target-driven incentive structures is creating stressful and competitive work environments. Many faculty, students, and staff in higher education also experience economic precarity or job insecurity; these working conditions constrain teaching, learning, research, and academic inquiry (Urai and Kelly 2023). The constant pressure to recruit more students, solicit more external funds, write more research proposals, and teach more classes has resulted in exhausted, burned-out academics who increasingly face mental health challenges (Hall 2023). The emphasis on individual performance and metrics leaves little room for collaboration and collective action. When the individuals learning and working in higher education institutions are struggling, a sense of purposeful commitment to the common good gets lost in both theory and practice.

Changing Direction

Global trends in each of the critical dimensions mentioned here are moving in the wrong direction. Although some universities are making changes to become more gender accepting and racially inclusive, political polarization and financial pressures in many places are countering these efforts. Without radical transformation, it is likely that institutions of higher education will continue to be complicit in perpetuating injustices, inequities, and disparities around the world. Confronting the injustices of higher education requires a paradigm shift, and the idea of climate justice universities provides a guiding framework for reimagining and envisioning what kind of transformations in higher education systems are possible.

Complacency is perhaps the most distinctive characteristic of hegemonic systems—individuals and institutions become accustomed to the ways of the system, accepting injustices and giving their attention, interest, concern, and inquiry to other areas. In many different ways, powerful interests that are resisting transformative climate justice are strategically promoting complacency to maintain the status quo. Universities are the institutions that should and could be counteracting the pressure toward complacency, yet all too often, higher education is being used to strengthen and perpetuate complacency. But the currently intersecting crises facing humanity are demanding that universities engage in new ways. The time is ripe to accelerate disruption of the patriarchal, white supremacist, colonial, and capitalistic structures that are fueling climate injustices, undermining well-being, and constraining the kind of societal impact that universities can have. Structural change is urgently needed not only to reduce the damage of current higher education systems but also to leverage the potential of universities to advance the common good and create a better future for all.

I began this chapter by sharing my personal experience of sexual violence in academia to demonstrate two things: how university hierarchies reinforce patriarchal oppression and abuses of power, and why transparency and accountability are essential for justice, equity, and inclusion to prevail. Subsequent chapters build on these critiques to reimagine climate justice universities by exploring alternative models and inspiring examples. Climate justice universities need to be committed to self-reflection, transparency, and accountability. For higher education institutions to advance a more just, equitable, and climate-stable future for all, a paradigm shift is needed to reclaim universities as organizations actively committed to co-designing and co-creating the redistribution and regeneration of knowledge, wealth, and power.

CHAPTER 3

Unlearning for Transformative Climate Justice

At the end of the spring semester in 2017, a student who had been in my Energy Democracy and Climate Justice course came to my office to thank me for offering the course. With generosity and enthusiasm, he told me that before he took my course he was like a fish swimming round and round in a fish bowl. But, he said graciously, the course experience had poured him into the ocean, and now he will be swimming freely in the open seas for the rest of his life. In all my years of teaching, the metaphorical gift I received from this student is a highlight. This vision of a student being liberated into the sea has nurtured and inspired me ever since. In this simple but powerful comment, I now realize that this student was describing the transformative power of unlearning. In the course, he had unlearned frameworks that were constraining his understanding of the world.

Unlearning describes the process of letting go of existing and constraining knowledge, beliefs, behaviors, and assumptions to allow for appreciation of new perspectives and information that may not align with our previous understanding. This concept has been defined in neuroscience in terms of how the synapses in our brains respond to fear and trauma (Marks and Tobeña 1990; Clem and Schiller 2016); theologically in terms of converting from one religious belief system to another (Michael and Wilson 2021); and in terms of sustainability transitions when individuals change their everyday practices (van Oers et al. 2023). Unlearning is a valuable concept when considering a paradigm shift in the societal role of universities because transformation requires more than moving toward something new and different. Transformation also requires intentionally letting go of perspectives and practices that are constraining us or no longer serving us. To consider whether current higher education systems are "fit for purpose" in the increasingly climate-destabilized world, unlearning is an important concept.

Given that humanity is stuck in a cycle of disconnection, precarity, and destruction, unlearning is essential to free ourselves from the shackles that we have created. Unlearning describes the liberatory power of knowledge. To swim freely in the expansive ocean of life, to have agency and power in the direction that we swim, requires a collective letting go of the constraints of a small fishbowl.

This chapter invites readers to playfully experiment with the idea of unlearning for both individuals (students and those teaching) and the ways universities structure their knowledge dissemination processes. I propose unlearning as a concept with value in reimagining the role of higher education in society because it provides a simple justification for why universities should let go of their conventional ways of curating and organizing knowledge. Most importantly, unlearning is a concept that encourages humility within higher education. Promoting and legitimizing unlearning within universities works to reduce the arrogance and false sense of certainty that is often projected from those academic experts who claim to already know the best path forward. Higher education institutions have a responsibility to stop reinforcing narrow, distorted perceptions of how the world works and instead nurture creativity and well-being beyond conventional economic measures of students' individual, competitive success. As we reimagine a paradigm shift toward climate justice universities, unleashing students' collective imaginations about what is possible not just for themselves individually in their own lives but for their communities and for other communities around the world could become a core empowering mission of higher education.

With the supposed democratization of knowledge through the internet and artificial intelligence (AI), the model of educational institutions as places where experts communicate knowledge to uninformed students is becoming increasingly outdated. Although many university instructors may still see themselves as the conventional "sage on the stage," expanded access to information means that the university's knowledge dissemination role is much broader than simply conveying information to students. Critical interpretation of different kinds of information is among the most fundamental skills of the twenty-first century. The rise of AI, including the emergence and accessible use of ChatGPT in 2023, has triggered widespread debate and speculation about how higher education will continue to adapt and respond (Aoun 2024).

Among the many academics who have reflected on how AI is impacting learning, literary scholar, neuroscientist, and university administrator G. Gabrielle Starr points out that it is the pleasure of the learning process—not the information itself—that leads us toward creative possibilities and active experimentation (Starr 2023). When that perspective is embraced, AI provides another tool to be used in the processes of both learning and unlearning. Given the numerous ways that AI reinforces racist and patriarchal biases and distorts human beliefs (Kidd and Birhane 2023), the concept of unlearning is particularly salient to considering the influence of AI on education. As higher education adapts and adjusts to technological advances in how knowledge is accessed and interpreted, intentionally engaging with the concept of unlearning within universities contributes to helping students make sense of the injustices of our time.

Around the world, students are struggling to understand the many paradoxes of climate injustice. Students see collective expressions of deep concern coming from global governance bodies, including the United Nations and the Intergovernmental Panel on Climate Change (IPCC 2023), but—at the same time—they see entrenched complacency demonstrated by an inability for powerful fossil fuel interests and mainstream institutions to implement the systemic changes that are so desperately needed. Students see the steady rise in climate vulnerabilities, including deadly heat waves, uncontrollable fires, devastating storms, and rising sea level, but then they also see a lack of action and commitment to change. These paradoxes cause anxiety, fear, and mistrust, and they also cause confusion, anger, and a sense of disempowerment (Servant-Miklos 2024). While educational systems cannot reconcile all of these contradictions, universities have a responsibility for providing students opportunities to explore a diversity of perspectives, to interrogate multiple kinds of power structures, and to examine alternative socioeconomic systems.

In this paradoxical era of contradictions and growing polarization, unlearning is a necessary part of opening up space to explore and reimagine more just, healthy, and stable alternative futures. By leveraging the idea of unlearning as a mechanism for change in how we think about learning processes in higher education, unlearning can be considered an essential part of encouraging counter-hegemonic thinking and action. Unlearning is part of a resistance to the dominant knowledge frameworks that are contributing to our ineffective and inadequate responses to interconnected crises. The concept of unlearning can facilitate the reclaiming of universities as sites for collective purpose toward the common good by a restructuring of higher education to include a more diverse plurality of knowledge systems that helps students understand the complexity of the world. Recognizing the disruptive nature of this concept, exploring the possibilities of unlearning within the university context is a way to liberate educational institutions from the tethers of past ways of thinking and anchor universities more strongly in the present with greater possibility for protecting the future.

The academic field of socio-technical transitions has explored how large-scale societal change requires both innovative ideas and experimentation, as well as simultaneous decline of the prevailing regime (Turnheim and Geels 2013). Resistance to change within the mainstream is inevitable and predictable because processes exist within institutions and structures to reinforce and perpetuate the status quo, mainstream regime (Geels 2014). To facilitate transformative change, therefore, educational institutions need to intentionally support both learning and unlearning. To address the proliferation of ecological destruction and the expansion of human suffering, higher education institutions could be creating more space for unlearning the knowledge frameworks and entrenched assumptions that are perpetuating crises. To actively confront the interconnected uncertainties, complexities, and ambiguities of our current predicament, it is necessary to unlearn the long legacy of disciplinary-specific knowledge dissemination practices that have been at the core of universities for centuries.

Proposing unlearning in universities is a provocative idea that will inevitably receive strong resistance. Many people perceive knowledge as cumulative so that we can scaffold new knowledge on top of existing knowledge. But the proposition here is that some knowledge needs to be intentionally deconstructed in order to make space for new and different ways of knowing. When we swim round and round in a fishbowl, we will not be able to experience the ocean until the glass is broken and the water is allowed to flow out.

I anticipate that the concept of unlearning may not align with some people's educational experiences. I also know this concept will resonate strongly with others. My hope is that this provocation to focus on unlearning will open up new space for different ways of thinking about how universities are engaging with, and responding to, our rapidly changing world.

To explore unlearning and learning to advance transformation toward climate justice, this chapter reviews alternative curricula, pedagogies, and epistemologies with a goal of inspiring structural and systemic change in what, how, and when teaching, learning, and unlearning occur in higher education institutions. *Curriculum* is the word generally used to refer to what we are taught, *pedagogy* refers to how we are taught, and *epistemology* refers to theories and frameworks for understanding knowledge itself. This chapter is the first of two consecutive chapters that explore reclaiming and restructuring knowledge to disrupt the conventional links among knowledge, wealth, and power in higher education. This chapter focuses on reimagining how universities disseminate knowledge through teaching, learning, and unlearning, while chapter 4 focuses on reimagining how universities generate knowledge through research, innovation, and exnovation.

Unlearning in Auroville, Tamil Nadu, India

Despite my extensive experiences in academic institutions of learning, it was not until 2018 that I fully appreciated the powerful concept of "unlearning." At this time, my dear friend from Argentina, Mariu Hernandez, whom I know from our years together studying environmental science in graduate school in Pasadena, California, was living in Auroville—the longest-standing intentional community in the world in Tamil Nadu in southern India. Mariu invited me to join a six-day convening event to celebrate and honor fifty years of the Auroville community. When this opportunity emerged, I was a tenured professor at Northeastern University in Boston and a single mother with my eldest, Cecelia, studying in university and my youngest, Aden, still in high school and living with me. I was intrigued by the invitation and eager to learn more about this unique collective social experiment, so I invited 16-year-old Aden to make the journey with me.

Auroville is an international experimental township made up of about 2,700 members from over fifty nationalities who are collectively committed to practical experimentation and research into transformation of cultural, social, environmental, and sustainable living (Clarence-Smith 2015; Clarence-Smith 2023). The community, which explores the spiritual needs for the evolution of humanity, was founded in 1968 on the teachings of Sri Aurobindo—renowned Indian yogi, revolutionary, and poet—by his spiritual counterpart, Mirra Alfassa, who is affectionately referred to within the community as The Mother. February 28, 2018, was Auroville's fiftieth anniversary. We joined the weeklong celebration to collectively appreciate how the community has developed innovative and alternative forms and practices that inspire people and projects worldwide. Auroville is a prefigurative community, meaning a community that lives the way it wants the rest of the world to be (Clarence-Smith and Monticelli 2022; Clarence-Smith 2023). Life in Auroville is considered a constant experience of experimentation, learning, and unlearning.

My experience of intentional unlearning during that trip was intense, inspiring, and joyful. Being able to share this experience of unlearning with teenage Aden—who is one of the most influential and wise teachers in my life—was humbling and powerful. The program for the fiftieth anniversary convening was full of demonstrations and reflections on alternative ways of living and thinking (table 1).

Table 1 Program for Auroville Becoming 50

Day	Theme	Focus	Guiding text	Source of guided text	Topics
1	Conscious collectivity	Experiments and practices that foster collective consciousness	"Auroville will be a site of material and spiritual researches for a living embodiment of an actual human unity."	Auroville charter	Solar kitchen, free store, universal basic income, learning societies, water management
2	Harmony from diversity	Approaching a lasting balance of individual, social, and environmental systems	"In this harmony between our unity and our diversity lies the secret of life."	Sri Aurobindo, *The Ideal of Human Unity*	Forest, energy, art, harmony, resilience
3	Unending education	Development in all fields of life	"Auroville will be the place of an unending education, of constant progress, and a youth that never ages."	Auroville charter	Awareness through the body, educational initiatives, yoga at work, transdisciplinary degrees
4	Creative progress	Diverse artistic expressions and cultural practices	"Make of beauty your constant ideal. Beauty of soil, beauty of thought, beauty of feelings, beauty of action, beauty in work."	The Mother	Art exhibits, theater for trauma, art education, spirituality in music
5	Collaborations	Connections	"Auroville wants to be the bridge between the past and the future. Taking advantage of all discoveries from without and from within, Auroville will boldly spring toward future realizations."	Auroville charter	Collaborative workshops to explore connections and potential new collaborations
6	Auroville's 50th Anniversary	Bonfire and water ceremony	Multiple	Multiple	Storytelling

Note: These events took place in Auroville, Tamil Nadu, India, from February 22 to 27, 2018, under the title The Bridge: Auroville and the World; A Collaborative Research Encounter in Auroville.

During the six-day event, we unlearned basic societal economic assumptions by living in a community with no money and visiting the community's free store—a store where everything is free and people are asked to take what they need and give what they can. We unlearned assumptions about individuals cooking for themselves by eating food prepared in a community solar kitchen. We unlearned assumptions about clothing by attending the annual "Trashion Show," which was a spectacular evening community event in which young and old danced across the stage of a large outdoor amphitheater to music, wearing exquisite costumes created out of trash. We unlearned assumptions about exercise by practicing yoga and participating in an "awareness through the body" experience. We unlearned our Western practice of eating with a knife and fork, using our right hand to scoop the local vegetarian curries into our mouths with roti flatbread. And perhaps most importantly, we unlearned assumptions about schools and formal education by experiencing a collective relational education based on the practical actions necessary for living a fulfilling life in community with others (figure 4).

Of the many people we met in Auroville, Manish Jain, an advocate of unschooling and unlearning, was among the most memorable. Manish shared with us inspiring stories about all the local knowledge and wisdom that he learned from his illiterate grandmother. He also told us devastating stories about how rigid formal education systems disempower, humiliate, and tear down so many young people who learn from an early age that they are inadequate and inferior. Manish advocates

Figure 4 (*opposite*) Photos from Auroville, Tamil Nadu, India, in February 2018. *Top:* Group convening during the weeklong fiftieth anniversary event. *Bottom:* The Auroville store, where no money is exchanged and people take what they need.

for reconceptualizing education as inclusive of every human being and every human experience. This requires unlearning hierarchies of knowledge, unlearning hierarchies among people, and unlearning hierarchies among educational institutions.

Manish is the cofounder of the Ecoversities Alliance, a global networked collective of learners and communities reclaiming diverse knowledges, relationships, and imaginations to design new approaches to higher education. The Ecoversities Alliance is a collective united by those exploring what the university might look like if it were at the service of our diverse ecologies, cultures, economies, spiritualities, and life within our planetary home. More details about the potential of expanding ecoversities—initiatives seeking to transform the unsustainable and unjust economic, political, and social systems/mindsets that currently dominate global societies—are explored in chapter 6 on local empowerment and global solidarity.

Unlearning Curricula of Deception

Curriculum refers to the content of what is taught in formal education institutions, including the ideas, concepts, and information that students are expected to learn. The current mainstream curriculum in most universities around the world is perpetuating a narrowly defined set of knowledge that those "educated" in certain areas of study should understand. This narrowly defined curriculum is what Manish Jain calls a "monoculture of the mind." Not only is the prescribed content of conventional university curricula narrow in scope, it also represents a biased view of humanity and a distorted perception of humans' temporal, spatial, and material relationships with the nonhuman parts of the world. Conventional disciplines that provide the structure for most university curricula are defined by a legacy of research and thinking based on narrowly defined histories and geographies. Because the curriculum in most universities does not prioritize exploration of the social systems, structures, policies, and practices that are perpetuating intersecting crises, universities are inadvertently encouraging most students to ignore, dismiss, or deny—rather than confront—the scale of current human suffering and ecological destruction.

The focus within emerging climate change education curricula is largely raising awareness of the changes that are happening rather than interrogating the socioeconomic systems that are preventing an adequate response (Fernandez, Thi, and Shaw 2014; Waldron et al. 2020). The growing contradictions that students recognize as they compare what they are learning in formal education with what they are experiencing in their lives is contributing to growing anxiety (Hickman et al. 2021). Given the complexities of understanding accelerating climate disruptions, curricula that avoid and minimize the climate crisis or curricula that offer certainty and simplicity are not just a disservice to students and society, but they are increasingly being characterized as curricula of deception (Matthews 2021). Introducing the idea of unlearning into university curricula provides a mechanism for challenging curricula of deception.

Disciplinary boundaries are intended to constrain and focus students' learning on specific kinds of knowledge. Students generally select their disciplinary focus, and then they are funneled into a curricular path that limits the breadth and diversity of knowledge that they will explore at university. Because of disciplinary constraints and the limits of current university curricula, there are big gaps in knowledge that constrain both individual and collective thinking about how to respond to climate injustices. For example, many students do not learn relational knowledge, that is, how humans are interconnected to earth's systems and the nonhuman world, or different alternative models of how societies and economies could be structured. Most students in universities do not learn practices of care, maintenance, and reciprocity, and there is often no space for lived experiences, embodied experiences, or deep reflection on the self (Facer 2019). In many university curricula, students are not exposed to knowledge about the interconnections between health and well-being for people and the planet, nor do they spend much time considering how social change happens.

The past twenty to thirty years have witnessed an expansion of curriculum on sustainability and climate change in education at all levels (Kelly et al. 2022). In higher education, the integration of sustainability into the curricula has evolved gradually from isolated specific courses and fields to a more interdisciplinary and holistic approach (UNESCO 2022). Initial courses began being offered in some universities in the 1960s and 1970s in response to growing environmental concerns. In the 1980s and 1990s, new multidisciplinary degree programs in environmental studies emerged in universities in the United States, Canada, and other places, and the 2000s and 2010s saw the establishment of many centers and institutes dedicated to promoting sustainability efforts in academia around the world; this resulted in the development of many more sustainability-related courses. More recently, efforts to integrate sustainability education across various disciplines have further expanded curricular offerings throughout the university (Sterling and Thomas 2006). Frequent justifications for the intentional expansion of these efforts include the societal need to train a future workforce with literacy in sustainability and also to build capacity to achieve the Sustainable Development Goals (SDGs) (Steele and Rickards 2021). Some universities have used the United Nations' SDGs, adopted in 2015, to guide sustainability efforts in attempts to align their curricula with the SDGs.

Research analyzing drivers for, and barriers to, integrating sustainability into the higher education curriculum around the world reveals four distinct types of integration: denial, bolt-on, build-in, and redesign (Weiss et al. 2021; Sterling and Thomas 2006). In this framework, "denial" refers to those universities that have implemented no change in the curriculum, while "bolt-on" refers to the inclusion of some isolated and discrete teaching about sustainability that is simplistic and instrumental, identified as first-order learning. The category of "build-in" refers to educational initiatives that are designed to advance sustainability—going beyond just teaching about sustainability. "Redesign" refers to transformative change anchoring sustainability at the core of higher education institutions, extending beyond education into all domains of the university (Weiss et al. 2021). Analysis of over 133 universities found that multiple factors determine the degree of curricular integration; these include strong institutional leadership, incentives, institutional commitments to advancing sustainability in research and campus operations, and professional development support (Weiss et al. 2021). In universities around the world, there is also resistance to integrating sustainability and climate change education into the curriculum. This means that some of the teaching and learning carried out in higher education institutions is contributing to climate obstructionism, slowing down social change and perpetuating climate injustice.

Teaching Regenerative Economics within Planetary Limits

The discipline of economics, and the associated field of finance, are arguably the academic areas where unlearning is most urgently needed to advance climate justice. In April 2023, the president of Ireland, Michael D. Higgins, gave a speech that generated significant defensiveness by mainstream economists in Ireland and beyond. As he spoke at a reception honoring the Think-tank for Action on Social Change (TASC), an Irish organization committed to research and public engagement on inequality, democracy, and climate justice, he made the claim that "a fixation on a narrowly defined efficiency, productivity, perpetual growth has resulted in a discipline that has become blinkered to the ecological challenge—the ecological catastrophe—we now face." He went on to say that the

> failure to facilitate a pluralism of approaches in teaching economics is a deprivation of basic students' rights, indeed citizen rights leading... to a narrow, blinkered and distorted education. ... Students are entitled not only to pluralism and the declaration as to assumptions of competing models in what is taught, but to be able to find intellectual and practical fulfilment in the engagement with ideas, ideas that will in turn be an influence on the options in advocated policy and their life contribution.

In this controversial speech, Higgins was referring to the fact that most students who take economics courses in higher education institutions around the world are taught a narrow, constrained view of economic structures and perspectives. The assumption that economic growth and free markets are essential for prosperity is often presented as an uncontested fact, and the physical and material boundaries of earth's finite systems are not always acknowledged. Although the assumption of infinite economic growth is a physical impossibility (Daly 2015), it forms the basis of most teaching in economics, finance, and business. Market fundamentalism, an ideology that prioritizes unregulated free markets, is pervasive in economic teaching, resulting from a long history of US industry investing in university programs in economics, business, and government to legitimize their resistance to government regulation (Oreskes and Conway 2023). This narrow and impractical approach to how economics is being taught in most universities has been described as a distinct form of deception (Erickson 2022). In his speech, the president of Ireland not only pointed out that students are learning a narrow framework that is disconnected from the realities of the world, but he also highlighted the distorted and destructive priorities that this kind of economics education perpetuates in society.

Calls for a radical change in the teaching of Economics 101 are growing (Røpke 2020). The majority of mainstream economists do believe economic growth is a requirement for improving the lives of the billions of people living in poverty (Pritchett and Lewis 2022), but within the economics community, a growing number of heterodox economists, including ecological economists and proponents of feminist economics, have been calling out what ecological economist Jon Erickson calls "the fairy tale of economics" (Erickson 2022).

The earth is a finite system with distinct material boundaries and absolute limits (Rockström et al. 2023; Steffen et al. 2015). Yet most students graduate from university without understanding these limits and without exploring how human socioeconomic systems are destabilizing the earth's regenerative and renewables systems. The focus in economics on continued economic growth—assuming perpetual material extraction without considering physical limits—is deceptive as well as disempowering to students (Stephens 2023). As climate change and ecological destruction accelerate, it becomes an increasingly critical responsibility for universities to teach students the importance of living within planetary limits.

The economic models and aggregation of economic data that many economists rely on do not include the physical limits of the earth's systems. Students have a right to learn that the earth is finite with multiple planetary boundaries (Steffen et al. 2015) and that regenerative practices and renewable resources expand the potential of a future of abundance, sufficiency, and health for both people and the planet. The planetary limits and the regenerative capacity of earth's systems are both fundamental to considering humanity's interconnected relationships with the nonhuman parts of the world. Yet the curricula used in teaching mainstream economics fails to adequately incorporate these concepts.

Another danger of mainstream neoliberal economics is that this kind of economics is often used to inform policy. At an event highlighting the importance of economics education hosted by the US Federal Reserve in October 2023, the chair, Jerome Powell, said, "Economics is the science of policy" (Powell 2023). Public policies informed by economic analysis that assumes infinite growth without acknowledging the earth's planetary limits are harmful to the health of people and the planet.

In university courses on economic, political, and social systems, students need to learn that the structure of the economy can either enhance or erode public well-being; conventional economic analysis provides only one lens to assess well-being. Students deserve to know that a plurality of economic theories and perspectives exist, and alternative economic systems are possible. Instead of continuing to reinforce that the current competitive and exploitative economic system is the only possibility, the climate crisis demands that mainstream economic assumptions are unlearned and alternative cooperative and regenerative economic systems are imagined.

Economists sometimes categorize themselves as either orthodox, which refers to mainstream, neoclassical, market-based thinking, or heterodox, which includes a plurality of alternative ways of thinking about economic systems, including feminist economics (Tejani 2019; Nelson 1996), ecological economics (Kallis and Norgaard 2010; Pirgmaier and Steinberger 2019), post-Keynesian economics (Fontana and Sawyer 2016), and other nonorthodox approaches. Diversity of economic thinking is expanding rapidly as the limiting dangers of dominant neoliberal economics become clear to more people.

During my time as a professor at the University of Vermont, I was immersed in the conflicting worldviews among economists that resulted in curricular segregation that I had never seen before. As a faculty affiliate of the Gund Institute, an institute founded on ecological economics and the need to disrupt GDP (gross domestic product) as the dominant measure of a nation's economy, I learned that the orthodox economics department would not allow students who were majoring in economics to count their "ecological economics" courses toward the economics major. So deep was the animosity and disrespect of the ecological economists who questioned the logics of mainstream economics that the faculty in the mainstream economics department saw the courses in ecological economics as outside of their discipline—not related or relevant to completing a degree in economics. This narrowness of inquiry and this exclusion of plurality in economics education is occurring in universities all around the world. To counter it, an urgent priority for higher education leadership must be to unlearn neoliberal economics and foster diversity in economic thought and economic curricula so that students can better understand the world around them. For students to engage genuinely with the social and ecological crises of our time, climate justice universities need to teach a plurality of economic theories and perspectives.

A British academic economist who developed a YouTube channel called Unlearning Economics (Unlearning Economics

2023) explains that he believes that many mainstream economists are actually unable to consider things outside the rigid theoretical way they were taught. He suggests that their minds have been unalterably limited by the old ideas. John Maynard Keynes, the British economist who developed ideas about the important role of government spending that were counter to mainstream economists, famously said in 1936, "The difficulty lies, not in the new ideas, but in escaping from the old ones" (Arun 1996). Arrogance, defensiveness, and an aggressive, male-dominated culture are additional challenges facing the many economists who are unable to consider realities outside of their simplistic and narrow worldview. Quantitative analysis of 460 economics seminars in 2019 shows that women presenting their economics research are treated with more hostility than men presenting their research (Dupas et al. 2021). This aggressive culture within economics has led to underrepresentation of women and others from systemically marginalized groups within the economics profession. In turn, this lack of diversity further constricts how the field of economics evolves and expands beyond the rigid, narrow dominant view.

Despite this (or maybe because of this), many of the most influential economists of the twenty-first century are women who are challenging mainstream economics by calling for radical transformation in the strategic role of the state. Here I refer to Kate Raworth and her call for doughnut economics, restructuring the economy within the constraints of planetary boundaries with a minimum basic economic support for everyone (Raworth 2017), and Mariana Mazzucato, whose work debunks the myth that the private sector is responsible for transformative innovation (Mazzucato 2013) and shows how public investments can change capitalism (Mazzucato 2020). Stephanie Kelton, a leader in modern monetary theory, is another influential economist who debunks the myth that a deficit is bad and calls for rebranding debt as a strategic investment in the future (Kelton 2020). Carlota Perez, who calls for a circular economy and redefining the "good life" with aspirational lifestyles (Perez 2008), is also challenging mainstream economists in impactful ways.

Interest in teaching and learning about alternative economic structures that prioritize well-being and ecological health is rapidly expanding around the world. One inspiring example is a degree program on Transformative and Sustainable Economies that was launched in 2023 by Lara Monticelli and colleagues at the Copenhagen Business School (Copenhagen Business School 2023). This program includes three required courses: Re-imagining Capitalism, The Political Corporation, and Organizing for Social and Environmental Change. It is designed to critically assess current political and economics situations and reimagine alternative ways of organizing and living. The program acknowledges that to change the status quo, a collective and ongoing reimagining of our economies, our societies, and the ways in which we organize and do business is needed.

Given the growing injustices and human suffering throughout the world, universities have a responsibility to introduce students to the limits of orthodox economics and encourage them to understand the potential of regenerative economics (United Frontline Table 2022; Fullerton 2015), evolutionary economics (Bergh et al. 2007; Foster and Metcalfe 2001), feminist economics (Oksala 2023), and other kinds of economics. Instead of focusing narrowly on corporations, business schools around the world could be teaching about cooperative models of enterprise, community ownership, and the possibilities of regenerative economics designed for well-being and prosperity (Hoffman 2021). Joan Robinson, a well-known British economist who contributed to post-Keynesian economic theory, said that "the purpose of studying economics is not to acquire a set of ready-made answers to economic questions, but to learn how to avoid being deceived by economists" (Erickson 2022).

Beyond economics, teaching about planetary limits or the climate crisis is not integrated into the curriculum at most universities; this means hundreds of millions of students complete a university degree without considering the ecological crises of this time. In part due to student demand, more universities are expanding their curriculum to include courses on planetary limits, climate destabilization, and ecological imbalance. In 2022, for example, the University of Barcelona, a university with over seven thousand students, became the first university in the world to announce all students would be required to take a mandatory course on climate change—a course that connects the social and ecological aspects of the climate crisis. This university-wide commitment came after student protesters from a coordinated End Fossil campaign occupied the campus for seven days. The students' argument was that it is irresponsible for any student to graduate without understanding the climate crisis and planetary limits (Burgen 2022).

Resources for teaching within planetary limits are accessible for instructors around the world. For example, the SDG Academy, the education and training division of the Sustainable Development Solutions Network (SDSN), a United Nations global initiative, promotes transformative education by creating and curating high-quality content on sustainability and sharing open access learning resources for a global audience (SDG Academy 2023). Another valuable resource, published in 2022 by the Club of Rome, a platform of thought leaders who identify holistic solutions to the world's most pressing global challenges, is *Earth for All: A Survival Guide for Humanity* (Dixson-Declève et al. 2022), which includes detailed analysis for a clear pathway to reboot the global economic system so it works for all people and the planet.

A free, open access *Climate Justice Instructional Toolkit* has been co-developed by environmental justice education expert Chris Rabe and students at MIT (Rabe et al. 2023). This resource, which is posted on the website of MIT's Environmental Solutions Initiative, provides teaching and learning resources for faculty, instructors, and students, including specific modules on climate justice fundamentals, climate justice policy, global climate justice, the just transition, indigenous climate action, mapping environmental justice, emotions of climate justice, energy justice, and mining and climate justice. This resource is designed primarily for undergraduate education, but an inclusive student-centered approach means this toolkit can be used for many different learning environments.

Unlearning Pedagogies of Oppression

Pedagogy describes how content is taught, that is, the approach the instructor uses to facilitate learning. Pedagogy includes how class time is structured and what kinds of assignments and expectations students are expected to engage with—for example, whether a course is based on lectures, projects, writing assignments, or exams. Cognitive research demonstrates that different people learn in different ways—so to create inclusive and accessible learning experiences, a diversity of pedagogical approaches is more likely to be effective than a single approach. While conventional pedagogical approaches assume that a professorial expert lectures to students who are attentively listening and absorbing what he is saying, interactive, engaged, and participatory pedagogies are more likely to support learning that inspires and empowers students to engage in social change.

In my own learning about pedagogy, I have been inspired

by two influential books: *Teaching to Transgress: Education as the Practice of Freedom* by bell hooks (1994) and *Pedagogies of the Oppressed* by Paulo Friere (1970). Both of these influential educators recognize the liberating potential of education while warning how education also serves as a mechanism to perpetuate systems of oppression. In this dynamic time of intersecting crises and climate disruption, the disempowering role of higher education teaching and learning in isolated classrooms with a "sage on the stage" communicating knowledge to a group of students is increasingly obvious. In response, many innovative examples of alternative pedagogies and resources are available to educators. One excellent resource of alternative pedagogies is *The Future Is Now: A Resource Catalog of Radical Pedagogies* by the Ecoversities Alliance (2020). Another is the praxis-oriented critical pedagogy of sustainability described by tina lynn evans in the 2012 book *Occupy Education*.

Unlearning specific pedagogical approaches is constrained in some places more than others. While there is extensive oversight, including standardization and regulation, in how specific courses are taught in many university systems in Europe and other places in the world, in the United States many instructors have had freedom to modify, adapt, and adjust their pedagogical approaches without seeking formal approval. This has allowed for creativity and experimentation in unlearning pedagogies of oppression (Casey 2017; evans 2012).

One example of an alternative university model with a unique pedagogical approach that resists pedagogies of oppression is the Campus de la Transition in France, an institution focused on training to transform by linking together ecology, economy, and humanism (Renouard et al. 2021). The pedagogy of the Campus de la Transition is based on discrete areas of teaching and learning that focus on living and discerning for social cohesion, interpreting and imagining, governing and

Table 2 Pedagogy of the Campus de la Transition in France: Six Discrete Areas of Teaching and Learning

Greek word	English meaning	Action for learning
Oikos	House	To live in a common world
Ethos	Behavior	To discern and decide for societal cohesion
Logos	Speech	To interpret, criticize, and imagine
Nomos	Law	To measure, regulate, and govern
Praxis	Action	To act in accordance with the stakes
Dynamis	Power	To reconnect with oneself, with others, and with nature

Source: Campus de la Transition, https://campus-transition.org/en/home/.

taking action, and strengthening connections (table 2). I have served on the Scientific Board of the Campus de la Transition since 2017, and throughout that time I have been impressed and inspired by the transformative pedagogical approaches they are implementing to achieve their goal of returning the economy to the service of humanity and nature. The campus, located in Forges in the Île-de-France region in north-central France, is a training center for social and ecological transition, a research laboratory, and a place for ecological experimentation for transformation toward a sustainable lifestyle.

The importance of unlearning pedagogies of oppression has been elevated with the rise of new AI tools, including ChatGPT. AI has forced change in pedagogical approaches in higher education as instructors adjust assignments to accommodate these powerful new tools that can instantly synthesize and organize information for learners. AI changes the ways that instructors facilitate learning, and widespread recognition of the dangerous biases of AI has resulted in a broad reassessment of how student work should be evaluated and what kind of assignments are most valuable in an era of AI. New pedagogical approaches, including more collaborative and creative assignments that focus on critical thinking skills and collective inquiry, are replacing more conventional pedagogical approaches of conveying information. Recognizing the negative impacts of the expanded reliance on technology for learning, there is also a growing movement in some universities to create technology-free zones. Some educators and students are calling for courses and spaces where students are required to spend time reading, thinking, and writing without technology (Worthen 2023).

Teaching Relational Knowledge and Reciprocity

Universities are traditionally structured to teach about humanity as if humans were separate and detached from one another and from the nonhuman world. The fragmented way that learning has been conventionally organized within institutions of higher education encourages students to conceptually separate themselves from others, the living from the nonliving, and the past and future (Kimmerer 2013). This separation creates the conditions for dehumanization, which encourages complicity and acceptance of violence, oppression, environmental degradation, genocide, and slavery. But we do not have to continue with this damaging convention. Instead of perpetuating disconnection and isolation, higher education institutions around the world can teach reciprocity by nurturing students' innate relational intelligence (Machado de Oliveira 2022). Understanding the complex interconnected linkages between social and ecological systems is a learning objective that could be central to every course, module, and degree program (evans 2012). There is no area of study that would not be strengthened by integrating this understanding.

One approach to teaching reciprocity is *fair trade learning*, an approach to education and experiential learning that emphasizes equity, reciprocity, social justice, and social responsibility. This pedagogical approach is inspired by the principles of fair trade commerce, which are designed to create more equitable and sustainable relationships between producers and consumers in global trade. Fair trade learning prioritizes ethical engagement with communities and tries to confront and address power imbalances and inequities often present in experiential learning, such as study abroad programs and community-based service learning projects (CBGLC 2023). Fair trade principles include ensuring mutual benefit from all parties involved, including the students, host communities, and local partners, and establishing respect and reciprocity by valuing local knowledge and experiences and mutual respect between learners and community members (CBGLC 2023).

Integrating the principles of *mutual aid* into pedagogical approaches provides another transformative approach to learning relational knowledge. With mutual aid, people are encouraged to "take what you need and give what you can." Rather than taking as much as you can (even if you do not need it), mutual aid recognizes that how much you take has an impact of what is available for others. There are multiple ways that universities can reinforce and demonstrate the cooperative and collective principles of mutual aid, including creating mutual aid groups among students (Molina and Jacinto 2015), facilitating mutual aid among faculty and staff (Bergart et al. 2023), and providing university-university support during times of crisis. An example of this is when some mainland US universities hosted scholars from the University of Puerto Rico in the aftermath of the devastating destruction of Hurricane Maria in 2017 (RISE 2023) (discussed more in chapter 6).

Given the global crisis of loneliness and the worsening sense of disconnection, universities need to do more to integrate the teaching of relational knowledge and care. Resisting and unlearning the individualistic mindset that is so common in competitive university settings, higher education institutions can restructure the way they facilitate learning to teach relational knowledge and demonstrate reciprocity. In her best-selling book, *Braiding Sweetgrass: Indigenous Wisdom, Scientific Knowledge, and the Teachings of Plants*, scholar and botanist Robin Wall Kimmerer explains that in indigenous culture an honorable harvest is defined as a harvest in which no one ever takes more than half (2013). If we recognize ourselves as part of a collective society in which everyone shares resources to meet their needs, honor requires us to anticipate the needs of others as well as our own needs and prioritize the opportunity and the responsibility to share the resources available.

Teaching Solidarity

In simple terms, solidarity is embodying ideas of unity, cooperation, and support for others, particularly for those who are facing social, economic, and political oppression. Solidarity can be taught, practiced, and promoted in many different ways in nearly all aspects of university activities and initiatives. But the disciplinary silos and output orientation of the contemporary university means that solidarity is often not centered or prioritized. Among the many challenging aspects of teaching solidarity in higher education institutions is confronting the frequency of times, places, ideas, and discussions that adopt an individualistic, isolated, disconnected, and complacent approach.

Despite the urgency for teaching and implementing solidarity as a fundamental response to intersecting planetary crises, solidarity is misunderstood, discredited, and often dismissed in many university curricula; teaching solidarity tends to be confined to specific areas of social justice and human rights (Eynaud and de Franca Filho 2023). But teaching itself can be an act of solidarity, and many scholars and scholar-activists around the world view their teaching as such (Manassah et al. 2022). One way to consider the value of teaching solidarity during this era of uncertainty and unpredictability is that although we have a limited capacity to see and know the future, as humans we do have a limitless capacity to care for it. A central part of our shared humanity is an abundance of care and compassion.

Unlearning Epistemological Hierarchies

All human cultures have knowledge systems and culturally specific scientific methodologies and approaches to understand the world. The diversity of global epistemologies is rich and expansive. Despite this diversity, higher education institutions prioritize what is often characterized as Western science, which refers to the scientific traditions and methodologies developed primarily in Europe and North America. Western science refers to a systematic approach to understanding the world by gathering and quantifying data and measurements, testing hypotheses and theories through experimentation with carefully controlled variables to isolate cause-and-effect relationships, and developing theories to explain and predict observed phenomena. Western science relies on reductionism, which is simplification of larger complex systems to understand an isolated part. Western science also strives for objectivity and impartiality, minimizing bias and subjectivity in observations and interpretation. This approach assumes that all people, if they are trained appropriately, will interpret scientific evidence in the same way.

The dominance of Western science perpetuated in universities around the world has resulted in the devastating loss of other ways of knowing. *Epistemicide*, a term that refers to the systematic destruction and erasure of knowledge systems, traditions, practices, and ways of thinking, draws attention to the harmful consequences of the dominance of Western knowledge systems on other cultures (Paraskeva 2017; de Sousa Santos 2008). This concept also highlights the connections among knowledge, wealth, and power and how the erasure of other forms of knowledge and the dominance of Western science has resulted in problematic unequal power dynamics that have resulted not only in the loss of biodiversity but also a global loss in a more inclusive and comprehensive understanding of the world.

In response, the drumbeat of calls for epistemic plurality is growing (Perry 2024). Argentinian scholar Walter Mignolo, a literature professor at Duke University exploring decoloniality, the geopolitics of knowledge, and pluriversality, advocates for epistemic disobedience to free ourselves and disentangle knowledge from geopolitics (Mignolo 2009).

Reparative Epistemic Justice

Although many universities around the world have been key actors in epistemicide, some are now recognizing a role and opportunity to contribute to reparative epistemic justice. Epistemic injustice occurs when different sources and types of knowledge are ignored, their credibility is questioned, or access to different types of knowledge, including traditional and indigenous knowledge, is blocked (Gupta et al. 2023). Epistemic justice acknowledges the loss of indigenous and local knowledges and recognizes that other racial and ethnic minorities, women, LGBTQ+ individuals, and others have been systemically marginalized and excluded from university knowledge systems. A long legacy of structural exclusion of certain kinds of people has prevented epistemic justice. Reparative epistemic justice tries to rectify this imbalance by acknowledging the ways in which certain kinds of knowledge have been marginalized. Educational institutions that engage in reparative epistemic justice are actively working to dismantle oppressive structures, centering marginalized voices, challenging dominant narratives, and fostering spaces for critical dialogue and reflection.

Examples of reparative epistemic justice include the establishment of departments and programs devoted to indigenous knowledge, Latina knowledge, Asian studies, African studies, and so forth. With growing recognition of the value and importance of indigenous knowledge systems, some higher education institutions have taken steps to incorporate indigenous knowledge into their curriculum. Some universities offer courses or programs specifically focused on indigenous studies, indigenous languages, and traditional ecological knowledge to provide understanding of indigenous cultures, histories, and ways of knowing. Some institutions have also established partnerships with indigenous communities to develop curriculum content that reflects indigenous perspectives. Given the low number of indigenous scholars in academic institutions around the world, collaborative relationships with indigenous elders, scholars, and community members have been important to inform how to incorporate indigenous knowledge into specific courses or across disciplines.

In her powerful and provocative book, *Hospicing Modernity*, Vanessa Machado de Oliveira describes the deep wisdom of indigenous knowledge systems as more mature and stable than the Western knowledge systems that are dominant in "modernity" (Machado de Oliveira 2022). She describes how indigenous knowledge systems are the great-aunt and uncle of Western knowledge systems watching as the young and reckless Western knowledge systems blow things up and cause a whirlwind of damage. We all need help grappling with the limits of these structures of modernity and understanding how these structures are gradually becoming obsolete, and we need frameworks for taking account of the "often invisibilized costs" of sustaining them (Machado de Oliveira 2022).

With regard to knowledge systems of late modernity, Machado de Oliveira describes the Brazilian saying that reminds us that in a flood situation we cannot learn to swim until the water reaches our hips. When the water is still at our ankles or our knees we can walk or wade through the water, and we may be able to see others already swimming in deeper water in the distance, but we can only swim once we have no other choice (2022). Before the water is deep enough, we develop ideas about how to swim and what we need to do to prepare to swim. These ideas are shaped by the level of water around us and our perception of how fast the water may be rising. In preparation, we can open ourselves to the teachings of the water and the teachings of those who have already been swimming through the flood. This metaphor reminds us of the powerful potential of universities prioritizing indigenous knowledge as well as the knowledge and experiences of other marginalized communities. Within the vision of reimagined climate justice universities, conventional knowledge hierarchies of individualism and disconnection would be dismantled and relational knowledge and indigenous wisdom would be recentered.

Unlearning as Liberation and Justice

As climate disruptions and ecological disasters become more frequent and intense, teaching students to think in ways that ignore, dismiss, or deny the climate crisis is no longer acceptable. Ignoring the climate crisis is not just an unfortunate omission; it is irresponsible, and it exacerbates and replicates the injustices of climate change. Nevertheless, efforts to restrict teaching and learning about the climate crisis and societal responses to climate change persist. In 2023, the state of Ohio proposed a law that attempted to outlaw the teaching of climate policy (Gearino 2023). Similar efforts to deny access to information about the climate crisis are emerging in other contexts, demonstrating the widely acknowledged power of learning and unlearning; powerful interests who feel threatened—not by climate change itself—but by proposed policy responses to addressing the climate crisis, are trying to constrain learning to hold on to their power.

Given the multiple injustices associated with current fossil fuel dominant energy systems, the need for unlearning in energy education in universities around the world is urgent. Students deserve not to be constrained in their understanding about energy futures and energy justice. Despite the rapid global expansion of renewable energy jobs, research shows that universities are failing to meet the growing demand for a clean energy workforce; 68% of the world's energy-related educational degrees still focus on fossil fuels while only 32% focus on renewable energy (Vakulchuk and Overland 2024). To advance liberation and climate justice, universities need to disassociate from fossil fuel interests to unleash unlearning in energy education and to dismantle the role of higher education in reinforcing fossil fuel reliance. To move society in the direction of climate justice and ecological well-being, higher education has a critical role in facilitating collective unlearning of the competitive, scarcity-based, fearful mindset that is perpetuating inequities and ecological destruction. Rather than continuing to invest in preserving disconnected and competitive learning environments in a world of increasing loneliness, isolation, and suffering, universities can instead offer students what so many people are desperately seeking—engaged connection, collective hope, and holistic transdisciplinary learning opportunities that empower and inspire justice, well-being, and solidarity (Kinol et al. 2023; Favretti 2023). A future with climate justice universities would harness the transformative power of learning and unlearning.

CHAPTER 4

Exnovation Research and Knowledge Co-creation

Throughout my US-based academic research training, I was mentored and supervised by a series of climate experts who genuinely believe that they are smarter than everyone else. These men—and the people I am describing here are all men—engage with a confidence that suggests that if only others understood the world the way they do then the climate crisis could be "solved." These climate experts with extensive knowledge in earth system science, physics, and engineering hold prestigious positions at elite institutions of higher education in the United States, and they leverage their power and influence to advocate for massive investments in technological fixes to the climate crisis. Unfortunately, the confidence they project as they meet with technology and finance billionaires and influential politicians, many of whom are eager to support concrete action on climate, reinforces a narrow focus on technological research and minimizes the potential for investing in social research to facilitate structural changes in economic and political systems. All too often, their arrogance also translates into toxic workplaces where condescension and intimidation are the norm. The hostile and competitive work environment discourages diversity of ideas because the intellectual contributions of those who have different perspectives are marginalized, and challenging the confident experts is very uncomfortable. This leads to a reinforcing cycle of arrogance because these experts are surrounded by subordinates who are carefully compliant.

I have always been troubled by the intellectual arrogance that is nurtured in the false meritocracy of universities. The competitive structures that define academic success and the hierarchal systems that determine research impact nurture a sense of superiority among some and a sense of inadequacy among others. The negative societal impacts of this arrogance do not only include devastating individual situations for those trying to navigate these hostile academic systems, but this arrogance also impacts what kind of research is conducted and who has the power to set the research agenda. When academic systems reward competitive, individualistic scholars and dismiss and disregard those with more holistic and collective priorities, disconnection and isolation define and frame university research.

One illustrative example of the dangerous outcomes of research advanced within a system that supports superstar technology researchers untrained in considering the power structures that disconnect them from the people and communities they claim their research could "help" is the rapidly expanding area of solar geoengineering research. Solar geoengineering refers to a proposed technical response to climate change that involves manipulating the earth's atmosphere to cool the earth by reducing incoming solar radiation. Sometimes known as solar radiation management (SRM) or climate engineering, solar geoengineering includes the idea of blocking incoming sunlight by spraying aerosols into the upper atmosphere. While this used to be a fringe idea, it has been strategically promoted by a small, but powerful, group of white, male scientists from elite universities in the United States. With a goal of reducing temperatures, this approach would require flying hundreds of specialized high-altitude airplanes continuously around the planet. This proposed approach comes with huge ecological threats for the planetary system, massive humanitarian risks for vulnerable people, and fundamental geopolitical challenges regarding who would manage and control its deployment. Despite these risks, multiple universities are accepting philanthropic support from wealthy donors to expand solar geoengineering research, and a strategic group of researchers is lobbying for public funding for solar geoengineering.

The small but influential group of scientists who have been advocating for more research on solar geoengineering has been successful in mainstreaming the idea (NASEM 2021). The ultimate "technical fix" (Weinberg 1967; Stephens and Markusson 2018; Markusson et al. 2017), this approach does nothing to address the cause of climate change, and the social, political, and ecological risks of advancing it cannot be understated (Stephens and Surprise 2020). Solar geoengineering technology creates a mechanism for powerful elites to manipulate the earth's climate system for their advantage. There is no way for this technology to benefit all regions of the world equally, so if and when it were deployed, some places may experience climate benefits, but other areas would be worse off.

Because the earth's systems are so complex and interconnected, attempts to manipulate the climate would have unpredictable cascading impacts. Among the many potential unintended consequences, changes to the earth's hydrological cycle could, for example, disrupt the monsoon season, which would threaten agriculture, reduce food production, and restrict drinking water access (Stephens et al. 2021). Another danger is that the prospect of this nontransformative technology could continue to slow down and detract from larger systemic changes to address climate injustices and human suffering.

Rather than promoting this kind of narrow technical innovation with a vast and diverse array of inevitable negative unintended consequences that could be devastating for so many, what if these university researchers were encouraged and supported to instead focus their attention and ingenuity on fossil fuel phaseout? Fossil fuel phaseout is an urgent policy priority and a climate strategy that has expansive and diverse positive co-benefits for human health and ecological health. But despite clear scientific evidence that fossil fuel production must be phased out to minimize climate chaos, the power and influence of those who are continuing to profit from fossil fuel production have effectively thwarted university research on fossil fuel phaseout. Global efforts to end fossil fuel production have so far been ineffective and insufficient in the face of the concentrated wealth and coordinated power of fossil fuel interests (Ahamed et al. 2024). Transdisciplinary engaged research about fossil fuel phaseout is urgently needed to inform policy, planning, and strategic engagement of global coordinated efforts to advance fossil fuel phaseout. But unfortunately, fossil fuel phaseout has been an intellectual "no-fly zone" for academic research due in large part to how research agendas are influenced by powerful corporate interests.

To open up our imaginations to explore a paradigm shift in the kind of research that could be conducted in universities around the world, we must collectively reflect on and understand the current state of academic knowledge production. In this chapter, I make the case that a different kind of research could be generated if universities prioritized and nurtured humility and a sense of collective interconnectedness instead of arrogance and a sense of individualistic isolation.

In his 1990 book exploring links among knowledge, wealth, and power, futurist and social critic Alvin Toffler makes the claim that knowledge is the most democratic source of power. At that time, Toffler was highlighting the transformative role of information and knowledge when access expands opportunities for individuals and communities to participate in decision-making and challenge existing power structures to shape the direction of society. But more than thirty years later, the rise of interest in solar geoengineering is just one example that demonstrates an alternative where narrow, expert knowledge can be leveraged to disempower people and reinforce authoritative rather than democratic power.

To ensure that knowledge generation in universities contributes to a liberating and democratic source of power rather than constricting and authoritative power, higher education needs to envision a restructuring of research practice and priorities. If academic research is to advance transformative climate justice and regenerative ecological health rather than create additional mechanisms for powerful people and organizations to reinforce the status quo by gaining more control over other people and the planet, the incentive structures and missions of academic research need to be reimagined.

This chapter explores new approaches to incentivizing, structuring, and supporting research that could result in research with a very different kind of societal impact. Rather than reinforcing the trend of university knowledge being leveraged to advance corporate interests and further accumulate wealth and power among privileged individuals and organizations, a different approach to supporting and guiding university research could result in a more transformative societal impact from academic research. If a more healthy, equitable, climate-stable future were a fundamental goal guiding university research, some research that is currently being supported would not be advanced and other research that is not currently being supported could be advanced. To leverage university research to support society-wide transformative change toward climate justice and move away from narrow, technocratic climate isolationism, I propose that two major shifts in research prioritization are fundamental.

First, the current research focus on innovation needs to be complemented with an equivalent focus on *exnovation*. While innovation refers to the processes of creating and implementing new or improved products, services, technologies, practices, and ideas, exnovation refers to the processes of eliminating or discontinuing products, services, technologies, practices, and ideas that were previously established or widely used and are no longer serving society well. Exnovation involves intentionally letting go of outdated or inefficient methods, technologies, or ideas to make space for alternatives. Just as universities need to focus on both learning new things and unlearning things that no longer serve us, we also need research that focuses on exnovation in addition to innovation (Davidson 2019). Exnovation requires recognizing the limitations or drawbacks of existing mainstream approaches and actively seeking to reduce and phase out certain practices and technologies.

Second, the mainstream practice of academic researchers conducting research without involving the people and communities impacted by the research needs to be reduced while the alternative research model of co-design and co-creation of knowledge needs to be supported and encouraged. The co-production of knowledge is an iterative, collaborative process involving diverse expertise and actors producing context-specific knowledge (Suresh Babu 2023). By co-designing and co-creating research priorities and research methods, this approach centers practical experience, resulting in more impactful research outputs with more relevance to societal needs. Knowledge co-creation is essential to bridging the knowledge to action gap, ensuring community relevance and applicability of research projects, and safeguarding against the arrogance of academic researchers who may unwittingly conduct research that is harmful, extractive, or dangerous.

Shifting Research from Climate Isolationism to Climate Justice

Since human-caused climate change was first identified in the 1970s and 1980s, climate research has focused primarily on science and technology. From those early days, the United States became a technology leader but policy laggard as it quickly became clear that investing in climate science and technology research was a prudent strategy because it did not require making other social or economic changes (Stephens 2009).

This technocratic approach reinforces climate isolationism, the term I have coined to describe the narrow way of framing climate change as a scientific problem requiring technological solutions (introduced and discussed in chapter 1). By framing climate change as an issue that is separate and disconnected from other social issues, climate isolationism upholds the status quo systems by encouraging nontransformative ideas to reduce the problem. By focusing climate investments on technological innovation with the potential to "solve" the climate crisis, this framing is based on patriarchal and colonial assumptions of control. Climate isolationism endorses climate policies that align with the commercial value of new technologies and the profit-seeking interests of those with power and influence. Climate isolationism is not only ineffective, exclusive, and nontransformative, it is also dangerous because all too often research attempts to treat the symptoms rather than the cause. Climate isolationism ignores issues of power, wealth, and growing inequities and in so doing ends up perpetuating and exacerbating economic precarity and climate injustices, which have compounding negative implications for democracy and social justice—and for the planet. Climate isolationism disempowers people and communities by narrowly focusing on technological change rather than transformative social change, which is essential for effective climate governance (see figure 1).

Climate justice, on the other hand, provides a more holistic and inclusive approach to research that requires a collective skepticism and resistance to narrow climate isolationism. A climate justice approach prioritizes university research that focuses on regenerative ecological and economic systems and how to restructure society to redistribute wealth and power rather than further concentrate it.

The growing interest and investment in solar geoengineering research provides an illustrative example of the dangerous implications of climate isolationism and how it contributes to further concentrating power among the hands of a few (Surprise 2020). Throughout the past decade, the United States has become a global leader in solar geoengineering research, with the largest solar geoengineering research program at Harvard University funded by philanthropic gifts from individuals and foundations, including Bill Gates (Stephens and Surprise 2020). In addition to contributing to delay and distraction from the structural and systemic changes that are desperately needed (McLaren 2016), solar geoengineering risks disrupting the earth's hydrologic systems, including altering the monsoon season in Southeast Asia, which could cause new regional disparities and injustices in food and water access (Abatayo et al. 2020), new global health disparities (Carlson et al. 2022), and further exacerbate biodiversity losses (Trisos et al. 2018). The imagined potential of solar geoengineering has created a new pathway for the rich and powerful to establish additional control over everybody else as climate impacts worsen (Stephens and Surprise 2020). The mainstreaming of solar geoengineering technology demonstrates the outsized social power of the polluter elite; a very few wealthy billionaires are driving the techno-climate conversation and perpetuating climate isolation and moving society further from climate justice.

As the climate crisis worsens, it is more important than ever to promote values-based research based on principles of humility and collective interconnectedness (Liboiron 2023). Perpetuating research that emerges from assumptions of patriarchal white male conceptions of privilege and power evolving from a colonizing and controlling mindset reinforces the systems and structures that result in climate injustices. If advancing climate justice were a core mission guiding university research, the research portfolios of institutions of higher education would look very different than they do now, and international climate policy might be more transformative. The time, money, and intellectual creativity currently invested in solar geoengineering research might instead be focused on fossil fuel phaseout, exnovating for a plastics-free future, or financial exnovation for climate justice.

The Urgent Need for Exnovation Research

In mainstream discourse, innovation is generally viewed as positive because it is associated with value and impact. Most people are unfamiliar with the term *exnovation*, and the "ex-" prefix has a negative connotation in the English language. The word *innovation* is usually associated with technological change rather than social change, and often, in many contexts, the negative impacts or unintended consequences and social harms of innovation are not acknowledged. With the global technology boom that expanded the internet, digital communication, and big data, technological innovation is increasingly considered a mechanism for strategic individuals and organizations to profit, and rarely are the negative distributional impacts of technological innovation considered. With a mainstream focus on innovation as a means toward individual and institutional success, investment in social innovation to advance the common good is often ignored and minimized. Similarly, the value of investing in exnovation is not yet widely recognized.

As humanity struggles to thrive in a world that has already overshot beyond multiple irreversible planetary limits (Raworth 2017), the need for research on how to dismantle, disrupt, and move away from harmful extractive and exploitative systems is urgent. A reframing is needed to ensure that higher education institutions can appropriately value the impact of investing in exnovation research. In this context, exnovation research can be considered intricately linked with a new kind of innovation that is explicitly decoupled from growth and capitalism. An international team of post-growth scholars has argued that the fundamental purpose of innovation should not be to increase productivity to contribute to economic growth (Robra et al. 2023). Rather, they point out that innovation should be defined as use-value creation, which means socially useful production that fulfills societal needs.

With the ubiquitous use of the word *innovation* in universities around the world, this shift in meaning allows for a different kind of research that explicitly includes both innovation and exnovation to fulfill societal needs. Valuing and investing in exnovation research is essential to move humanity away from the worst ecological disruptions. The societal impact of these kinds of exnovation research agendas have potential to be transformative on multiple different levels; research on global coordination of fossil fuel phaseout, for example, may impact international climate negotiations as well as local cities and towns making decisions on how to prioritize energy infrastructure. Without a shift to valuing and incentivizing exnovation research, the perceived infeasibility of fossil fuel phase-out is likely to remain.

To demonstrate the undervalued impact of exnovation research, consider research focused on phasing out the use of cancer-causing products and practices. The billions of dollars that are spent annually on developing advanced cancer treatments are valuable, but why are we not spending an equal amount on research on how to exnovate away from the cancer-causing products and practices that are in widespread use? The commercial value for developing potential cancer drugs and treatments is what has incentivized the research to focus on the cure rather than the root cause of cancer, but from a public health and social justice perspective, society should be investing much more in preventive cancer research.

Economic and Financial Exnovation

A key area in need of exnovation research is economics and finance. For example, interdisciplinary economic exnovation research could adopt the framework of doughnut economics, Kate Raworth's economic model that balances essential human needs with planetary boundaries. The doughnut represents the ecologically safe and socially just space where the economy should remain within two concentric rings: the inner ring is the social foundation to ensure no one is left falling short of life's essentials, and the outer ring represents the ecological ceiling that ensures that humanity does not expand beyond the planetary boundaries that protect earth's supporting systems (DEAL 2023). To co-develop policies, practices, and structural changes to remove the unrealistic growth imperative from current political and economic assumptions, doughnut economics exnovation research could be supported in universities around the world. Higher education could collaborate with local communities and organizations to contribute to shifting toward regenerative economic models. Universities could partner with Raworth's Doughnut Economics Action Lab, which provides tools and resources to translate the doughnut economics idea into transformative action at multiple scales (DEAL 2023).

Another exnovation research area that needs more attention is how to transition away from the currently dominant unrealistic economic assumptions of infinite growth. The basic ideas of doughnut economics and post-growth transformation are that growth is a phase that cannot continue forever, so given planetary constraints to growth, we have to develop a different paradigm rather than constantly striving for growth (Raworth 2017). We know from ecological realities that growth of living things, whether it be a tree or a human being, is a critical phase of development but it cannot go on forever. Once a certain maturity and size is achieved, the living system adjusts to prioritizing thriving and continuing to live without continuing to grow in size. This same logic can be applied to our economic systems, which in the twenty-first century need to adjust to prioritize thriving and continuing to persist without continuing to grow. The biggest challenge in advancing this post-growth paradigm shift is that nation-states, policymakers, municipalities, and enterprises of all kinds have come to rely on sustained economic growth as foundational. So, it is challenging to reimagine our policies and organizations without prioritizing growth. But it is possible, and research in this area is urgently needed (Hickel et al. 2022). Universities and academic researchers have a lot to contribute to the complicated process of moving away from the constant growth paradigm. In this context, exnovation research will facilitate the inevitable and messy path away from the unrealistic assumptions of infinite growth (Hickel et al. 2022).

Exnovation research on post-growth transformation has

begun, but so much more is needed. In 2022, the European Research Council Synergy Grant program allocated 9.9 million euros toward research on pathways toward post-growth scenarios (ERC 2023). At the time, this was the largest funding ever for degrowth research. This exnovation research involves developing a range of scenarios, creating new models to project human well-being, and exploring and describing what post-growth policies could look like. This research will develop and propose democratic models of provision systems to ensure future generations have adequate access to food, energy, shelter, health, and social security, and it will explore and identify specific political and practical steps toward this paradigm shift (ERC 2023).

Additional research exploring potential policies to cap income and/or wealth provides another example of a new kind of research that simultaneously considers the innovative potential of new economic policies while also exploring the exnovation required to increase political and public support for these policies that is counter-hegemonic, meaning it requires letting go of some widely held assumptions (François, Mertens de Wilmars, and Maréchal 2023). Despite growing awareness that reducing the income and assets of the wealthy must be included in any strategy to reduce inequity, research exploring caps on wealth and income is minimal to date, and this area represents an important emerging research agenda. The research conducted so far in this area points to historical examples where wealth cap policies have reduced economic inequity in the past and suggests how policymakers could draw on those examples to design new wealth and income cap policies that decrease inequities in addition to being widely popular (François, Mertens de Wilmars, and Maréchal 2023).

Another rapidly growing area of exnovation research involves exploration of various forms of reparations and reparative processes that attempt to compensate for some of the cumulative harm resulting from the legacy of structural and systemic injustice. Some examples of research in this space include exploration of returning land that was taken away from dispossessed Black families in the United States (Burch 2023), returning land to indigenous communities whose unceded land was taken by colonizers (Tuck and Yang 2012), analysis of integrating reparations into infrastructure planning (Song and Mizrahi 2023), and research on compensation for disproportionate climate impacts among those who have not contributed to the accumulation of emissions in the atmosphere (Fanning and Hickel 2023). As the injustices of climate disruptions expand and become more obvious to all, the case for climate reparations is growing (Táíwò and Cibralic 2021), and research exploring the responsibilities of fossil fuel companies to pay for climate damages is expanding (Grasso and Heede 2023).

These are just a few examples of exnovation research in economics and finance. Additional discussion of financial innovations for climate justice universities is included in chapter 5 on regenerative financial structure. As the limits to growth become more obvious and the economic costs of climate disruption expand, economic transformation of some kind is inevitable. Expanding economic exnovation research to inform alternative structures and policies to move away from current unsustainable economic systems will provide urgently needed analysis, vision, and hope.

Fossil Fuel and Plastics Exnovation

Research on how to phase out fossil fuels and how to move toward a plastics-free future are two other important areas of exnovation. If academic research were oriented toward climate justice with ambitious goals of addressing the most important challenges of our time, an expansive and inclusive interdisciplinary exnovation research agenda focused on fossil fuels and plastics would have already developed.

Although powerful interests have invested for decades in ways to resist policy and research focused on fossil fuel phaseout, this is starting to change. After decades of avoiding direct discussion of fossil fuels, in 2021, subsidies for fossil fuels were finally mentioned for the first time in international UN climate negotiations at COP 26 in Glasgow. A rapidly growing global network of policymakers, scholars, and activists is calling for a global fossil fuel nonproliferation treaty (Newell, van Asselt, and Daley 2022), and fossil free zones are emerging in communities and organizations (F. Green 2022), including at some universities (discussed more in chapter 6). In 2022, a UK Research and Innovation Frontier Research Grant (originally a European Research Council grant, but because of Brexit the United Kingdom ended up funding the project) was awarded to a research team at the University of Sussex to study supply-side national and international climate policies that leave fossil fuels in the ground (UKRI 2023). Other researchers have begun studying fossil fuel bans (Green 2018) and the norming involved in anti–fossil fuel activity (Fitzgerald 2023). While this kind of exnovation research around fossil fuel phaseout has expanded in the past few years, it is still a small academic area. More research on how to collectively advance fossil fuel phaseout is desperately needed in universities around the world.

After over a decade of conducting research on renewable energy and electric grid innovation (Stephens, Wilson, and Peterson 2015), I realized how little research was being done on how to accelerate and manage the phaseout of fossil fuel energy systems. So, at Northeastern University in 2017, I launched a collaborative research team of several faculty researchers and graduate students to focus on fossil fuel phaseout. Our research in this area included analyzing how large multinational fossil fuel companies are publicly communicating about the transition toward a renewables-based future (Si et al. 2023), exploring connections among the fossil fuel industry, the plastics industry, and the agrochemical industry (Kinol et al. in press), and understanding how fossil fuel interests influence universities and research agendas (Kinol et al. 2023). Our team also collaborated with colleagues from Puerto Rico to study fossil fuel obstruction and discourses of delay slowing the pace of renewable transition in Puerto Rico (Kuhl, Stephens, et al. 2024). In April 2023, our collaborative research team organized an interdisciplinary, open, free, public-facing conference on fossil fuel phaseout research at Northeastern's London campus. Given how central fossil fuel phaseout is to addressing the climate crisis, the novelty of research in this area is shocking and can only be explained by the powerful influence fossil fuel interests have in setting academic research agendas.

Another example of exnovation research related to fossil fuel phaseout is attribution science, meaning science that attributes climate impacts to specific fossil fuel companies. Research in this area is growing rapidly, providing evidence for court cases to hold specific companies and governments that subsidize and enable fossil fuel companies accountable for climate damages. As climate damages increase, communities, policymakers, and legal experts are increasingly asking who bears responsibility for the rapidly growing costs (UCS 2023). Attribution science research has been able to identify that 19.8 million acres of forest area burned by wildfires across western North America since 1986 is attributable to heat-trapping emissions traced to eighty-eight of the world's largest fossil fuel producers and cement manufacturers; these polluters contribute to nearly half of the increased fire-danger conditions across the region (UCS 2023).

How to phase out plastics is another critically important area of exnovation research that has not yet been prioritized in higher education despite the imminent risks to both human health and ecological health. Abundant physical science research documents the scale, scope, and health impacts of plastic pollution (Iroegbu et al. 2021; Blettler et al. 2018; Forrest et al. 2019), including studies showing microplastics have been found in both human breastmilk and cow's milk. Biological science research on the impacts of plastic pollution on biodiversity and marine species is growing (Bergmann et al. 2022; Teichert et al. 2021). Despite all the research on the harms of plastics, there is no large coordinated university-based research program focused on how to phase out the prolific use of plastics in contemporary society (Horton 2022).

The lack of attention to exnovation research on plastics follows a similar logic as that of fossil fuels; the plastics industry—which is directly linked to the fossil fuel industry because plastics require petroleum inputs—has strategically invested for decades to resist research and action that would reduce plastic demand. Instead, the industry continues to promote research that sustains and expands the global market for plastics. To distract us from the dangers of plastic pollution, the industry created an extensive public campaign to promote recycling, even though it has been revealed that plastic recycling has never worked in the ways industry claimed (Enck and Dell 2022). With increasing global concern about plastic pollution and its environmental and health impacts, exnovation research in this area is urgently needed, and universities have opportunities to partner with civil society, governments, and nongovernmental organizations working to phase out plastics. One inspiring organization in this space, A Plastic Planet, is collaborating with suppliers and designers to challenge the textile industry to create plastic-free fashion; alternatives to plastics are available, but "fast fashion" has gotten so accustomed to relying on cheap plastics that the transition away from plastics requires transformative changes in policy, practice, and incentive structures.

A healthy future for people and the planet requires a rapid phaseout of both fossil fuels and plastics. With a conceptual shift toward climate justice universities, we can envision alternative funding streams and incentive structures that support exnovation research in these specific areas of transformation.

Knowledge Co-creation and Regenerative Research

Moving away from technocratic, isolationist research requires a new model of regenerative, co-created research processes. The "knowledge to action gap" is a phrase used to describe how irrelevant and disconnected so much academic research is to the nonacademic world. With growing recognition of this gap, universities around the world have been publicly committing themselves to new forms of partnerships and community-engaged research throughout the past decade.

Knowledge co-creation refers to a collaborative creative process of research where a diversity of perspectives, experiences, and actors are included (Grindell et al. 2022). Regenerative research describes research that contributes to the regeneration of personal and planetary health within the "safe and just equitable space for humanity," that is, it is research that responds to the social and ecological distress that humanity is facing (van den Berg 2023). Most contemporary universities do not currently characterize, prioritize, or support knowledge co-production or regenerative research; a paradigm shift toward climate justice universities would encourage and support this kind of research.

For universities to be effectively leveraged for transformative change in society, knowledge co-creation needs to become the norm rather than the exception. As social infrastructure in many communities is being eroded—including the decline of local media and the closing of community centers for health, wellness, and local organizing—institutions of higher education have a new role to play as trusted sources of information and places for communities to convene. Corporate interests are strategically investing in misinformation campaigns, which is causing confusion, particularly in public health, about what is legitimate scientific knowledge and what is fake. Universities have a large role to play in stepping up to provide trustworthy resources for people and communities. Rather than focusing on producing research to be appreciated by other researchers or to be commercialized to make profit, university research should be reframed and realigned with community needs and well-being. Conducting research that is relevant to the experiences of neighboring and vulnerable communities in the region requires new modes of co-designing and co-producing research, including collaborations with community leaders and local health officials and coordination with local community organizations and social service agencies. This requires a mindset shift among university researchers to prioritize our human connections and our collective impact rather than prioritizing our individual intellectual and academic contributions.

One inspiring example of a new engaged institutional model of knowledge co-creation is the Dutch Research Institute for Transitions (DRIFT), which was a part of the Erasmus University Rotterdam until the university kicked it out because its approach was deemed nonacademic. DRIFT develops and shares transformative knowledge with community-based clients with a goal of accelerating transitions toward more just, sustainable, and resilient societies. In 2022, when the university realized that DRIFT was generating more research grants and research publications than other parts of the university, and the community-engaged, transition-focused approach was clearly having more societal impact than more mainstream research areas, the university invested in a new initiative to integrate this more engaged approach throughout the university (Wittmayer and Loorbach in progress). DRIFT is mentioned again in chapter 6 to demonstrate the potential of centering community in university activities and initiatives.

With community-engaged research, academics also must be cautious not to reinforce extractive and hierarchal dynamics in the relationships. Colleagues at Northeastern University co-developed with community partners a set of principles of anti-oppressive community engagement for university educators and researchers (Riccio, Mecagni, and Berkey 2022). These principles include honoring communities' autonomy and right to self-determination and respecting communities' history, culture, lived experience, and expertise. Recognizing the limits of our lived experience, expertise, and perspectives and reflecting on our social identities, positions, and power are also important principles. Building authentic, mutually beneficial relationships with patience and humility, managing resources equitably, and holding ourselves accountable to the values and principles of anti-oppressive community engagement are also included. Rethinking our relationship with time and urgency, and prioritizing patience, perspective-taking, and joy are additional principles of anti-oppressive community engagement. These principles are challenging to implement in most universities because of the structural power differentials but are necessary to facilitate meaningful collaborative, community-centered research.

Regenerative research is another key priority for reframing and restructuring the intentions, priorities, and practices of academic research. Regenerative research refers to research that contributes to the regeneration of personal and planetary health by relying on connections and cooperation rather than competition. Regenerative research builds capacity in the community, and it includes understanding and promoting local heritage, including the ecological, geological, biodiversity, archeological, and community dimensions of local communities. Opening up to regenerative research that relies on co-creation provides a mechanism for expanding the different types of knowledge that are generated and prioritized in academic research—including indigenous and local knowledge (Orlove et al. 2023). When regenerative research is prioritized within universities, expansive possibilities for transformative impact and community benefits emerge.

Resisting Corporate Influence and the Research-for-Profit Model

The current system of academic research is dominated by the quest of higher education institutions to bring in large external grants and to support research that could have commercial value. In this model, researchers who are successful in bringing in grant money and researchers who demonstrate potential for commercialization and private sector interest in their research sometimes become stars while those researchers doing more creative or disruptive work focused on the public good are often marginalized and disempowered.

Reimagining a different kind of research impact requires intentional resistance to corporate influence and the research-for-profit model that is so pervasive in the current research landscape. Multiple studies have demonstrated that industry sponsorship of university research has historically biased research in favor of numerous sponsoring industries, including tobacco, pharmaceutical, food, sugar, and lead (Fabbri et al. 2018; Legg, Hatchard, and Gilmore 2021; Morris and Jacquet 2024; Oreskes and Conway 2010). Systematic reviews of pharmaceutical industry–sponsored studies found them to be more favorable toward the sponsor's product than nonindustry-sponsored studies (Lundh et al. 2018).

In addition to the competitive, financial pressure for universities to secure external research funding, other adjacent sectors are profiting from the research activity at universities. Academic publishing extracts profits from the free labor of researchers who not only pay to publish but also provide their free labor to the profit-seeking publishing companies to conduct peer review. Resistance to this research-for-profit model has been growing. For example, in 2023, the entire editorial board of a journal published by Elsevier, the Dutch publishing company that publishes a large percentage of all academic research, resigned (Fazackerley 2023). The mass resignation was an act of protest and an expression of outrage regarding the level of profit disclosed by the publishing company.

To enable research for climate justice, strategies deployed by industry to deliberately manipulate university partnerships to further their own profit-seeking interests need to be exposed, revealed, and resisted (Franta 2021). A 1978 document written with advice for industries facing regulation, for instance, advised "coopting" academic experts and "identifying the leading experts in each relevant field and hiring them as consultants or advisors, or giving them research grants and the like . . . it must not be too blatant, for the experts themselves must not recognize that they have lost their objectivity" (Owen and Braeutigam 1978). A 1998 internal strategic memo from the American Petroleum Institute (API), which was leaked and is now publicly available on the Climate Files website, revealed the industry's strategy on "build[ing] a case against precipitous action on climate change" (American Petroleum Institute 1998). This document, which was developed soon after the Kyoto Protocol was signed in 1997, shows that API advised establishing "cooperative relationships with all major scientists whose research in this field supports our position" and organizing "campus/community workshops/debates on climate science" (American Petroleum Institute 1998). The goal was clear: the document says that "victory will be achieved when average citizens understand uncertainties in climate science" (American Petroleum Institute 1998). More specific revelations about the fossil fuel industry's strategy and its partnership with Princeton University are reflected in an internal 2017 campaign strategy memo presented by a public relations firm to fossil fuel giant BP. In this memo, the firm proposed targeting Princeton University as a "partner" helpful in "authenticating BP's commitment to low carbon" (Brunswick Group 2017). While fossil fuel industry contributions to US and Canadian universities have been well documented (Leonard 2019), contributions from the US-based Koch Foundation to several UK universities were revealed more recently, demonstrating the international networks and strategies of industry influence of universities (Colbert 2023). Recognizing how academic research has been captured by financialized commercial interests is a necessary first step in reimagining what a research system committed to climate justice might look like when it has strong ties to community needs rather than commercial interests.

Reclaiming Research for the Common Good

Rather than academia serving as gatekeepers defining for society what kind of knowledge is valuable (Haraway 1988), an alternative research system could be designed to incentivize research to address the structural social failures humanity is facing. While the training of medical doctors includes multiple reminders for doctors "to do no harm," there is no such framework in the training of university researchers. Academic research is inquiry based, with minimal reminders for researchers to consider the potential harm of their research. A widely held assumption of academic research is that all knowledge creation has societal value because it expands the knowledge horizon. There are minimal ethical guidelines or harm-reduction principles to guide university research. Rather, it is financial support for specific research agendas that is the biggest influencing factor in determining what research is conducted in universities.

With the rise of corporate and philanthropic funding for university research and growing political influence of powerful interests shaping the research agendas of publicly funded research, academic research has been increasingly oriented toward commercial value and is less focused on research questions most relevant to public health and the public good (Fabbri et al. 2018; Legg, Hatchard, and Gilmore 2021). This corporate influence over research agendas has been named "the Science for Profit Model" (Legg, Hatchard, and Gilmore 2021), and it represents knowledge generation being leveraged to further concentrate wealth and power. Controlling the research agenda not only allows powerful interests to advance knowledge that they may be able to profit from, but it also provides private industries with another mechanism to affect policymaking by influencing the type of evidence that is available, constricting the kinds of interventions to be considered (Fabbri et al. 2018), and legitimizing industry actors among academics, lawmakers, and the public (Legg, Hatchard, and Gilmore 2021).

In the medical context, the challenges of defining harm are widely acknowledged, and it is increasingly clear that understanding harm requires physicians to consider how treatment decisions are constrained by the patient's race, cultural background, and economic conditions (Sederstrom and Lasege

2022). One approach to reducing the complexity associated with defining harm in the medical profession involves promoting a holistic and preventive approach to care; Dr. Louis Lasagna, a physician at Johns Hopkins University, proposed in 1964 that all doctors pledge to "prevent disease whenever I can, for prevention is preferable to cure" (Sederstrom and Lasege 2022). A similar framework and mindset are needed in considering academic research. University researchers need to engage and collaborate directly with diverse communities so that research design is based on preventing harm rather than trying to fix harm after it has happened. Intentional recognition of the disparate needs of people with different cultural, racial, and economic experiences is essential for minimizing harmful impacts of research.

Within the traditional university research landscape, only a handful of brave researchers have openly reflected on the potential harm of mainstream research. The work of Eve Tuck, an indigenous academic, has pointed out the disempowerment that results from what she calls damage-centered research, that is, research that exposes and documents people's pain and brokenness (Tuck 2009). Researchers conducting research that exposes and reveals oppression and suffering may intend for their research to hold those in power accountable; they may hope that the exposure will prevent future harm, trigger reparations, or funnel resources for marginalized communities. But researchers do not always realize that this kind of research also reinforces the perception of those people as hopeless and depleted. With her collaborator K. Wayne Yang, who is a provost of University of California, San Diego, Tuck expands to describe why refusal to engage in this harm-centered research may be necessary to prevent social science research that disempowers, diminishes, and rehumiliates people (Tuck and Yang 2014).

The growing international network of scholar-activists called Faculty for a Future (also mentioned in the conclusion) provides a research toolkit, which is a resource for researchers who feel a responsibility for the work they do to limit harm from the multiple crises facing humanity (Faculty for a Future 2023). As part of its "people-powered universities campaign," this toolkit includes suggestions on ways to configure research to make a difference, advice on dealing with differences in status and access to power, and practical tools for trust-building between diverse perspectives, together with guidance for securing funding for this kind of work.

To contribute to the public good, those investing in and guiding university research must acknowledge that neutral knowledge does not exist. As Afro-feminist Sylvia Tamale describes in her 2022 book about unlearning imperial power relations, knowledge production, or what and how we understand "reality" and "truth," is an extremely political process. Research is not and has never been pure exploration and discovery; it is constrained by funding, and it is also constrained by assumptions about stability and different interpretations of stability for whom.

Research is an inherently social process. We accumulate knowledge, but when and how does that new knowledge change our practices in society? This is the crux of the challenge associated with structural change. Researchers are increasingly drawn to problems that funders have highlighted as important, so research agendas are prescribed not by demonstrated public need but by those who distribute research funding. A bias toward technological research and technological solutions to narrowly defined problems is reinforced by the reality that many researchers are not trained to focus on systemic change and the potential for transforming social structures. The prioritization of technological change research over social change research has minimized understanding of and appreciation for the possibilities for structural transformation.

Changing how researchers are trained is a challenge because research training is kind of like an apprenticeship. Students work closely with established researchers who guide the research project. The students gain valuable experience, often while contributing to the research agenda of their advisor. This model means that conventional approaches are passed along, and novel ways of conducting research may be slow to catch on.

Knowledge creation is a clear goal in universities; however, the question "knowledge for whom?" is rarely asked. If higher education is to contribute to reducing suffering and advancing human and ecological health, distributive and regenerative research, with a new focus on exnovation, needs to be a priority for academic research.

CHAPTER 5

Regenerative Financial Structures for Higher Education

Generosity and loyalty are among the defining characteristics of our alumni.

—Lawrence Bacow, president of Harvard University, in an email to the Harvard community announcing an unrestricted gift of $300 million from Kenneth Griffin, an alumni hedge fund billionaire (April 11, 2023)

In the middle of April each year, Northeastern University designates "Giving Day"—a specific day when everyone in the Northeastern community is asked to make a personal financial contribution to the university. Such an indiscriminate request in which all faculty, staff, students, and alumni, as well as families of students and alumni, are encouraged to act as philanthropists has become common practice at many higher education institutions. As universities recognize the wealth and disposable income among many individuals within their networks, some are constantly and strategically cultivating donors.

Each year, Giving Day at Northeastern made me feel uncomfortable. My discomfort emerged from the inherent power dynamics involved in this community-wide request to participate. The senior leadership, including my direct supervisor when I was working there, was involved in coordinating enthusiasm and normalization of this request for everyone in the university to donate money from their own bank account to support this rapidly growing private institution whose total assets increased to $5.5 billion in 2022 (net assets were $3.47 billion). Why should students—of whom more than 70% will graduate with thousands of dollars of student loans—be expected to donate financially to the university? That the request is directed at low-income staff and faculty on precarious, short-term contracts seems particularly problematic. This practice, which has become standard in many financialized universities in the United States, seems exploitative because of the power differential between employees who rely on the university for their livelihoods and their supervisors who pressure them to contribute their hard-earned money back to the university.

On Northeastern's Giving Day in 2023, I received over twenty different emails asking me to donate. I first received a series of emails from my direct supervisor, who was designated as one of the university's "Giving Day ambassadors"; I did not envy her situation as she was charged with leveraging her position of power to make the case to all the faculty and staff in our unit for why we should all "make an impact through philanthropy." I also received a series of email requests from the college-level team and another series of emails directly from the university's advancement office. Because, at the time, I was a parent of both a current student and a recently graduated alumni (Northeastern offers free tuition to all employees' dependents, a valuable benefit and a powerful retention strategy), I received another series of emails to families of students and alumni asking for financial contributions. The overwhelming number of email requests, the peer pressure, and the repetitive message within the barrage of requests that day felt coercive.

The discomfort I felt was similar to my reaction over twenty-five years earlier before my Harvard University undergraduate commencement in 1997. As preparations were being made for the graduation ceremonies (and my proud Irish grandmother was coming over from Dublin for the big event), I remember a few of my peers asking everyone in our graduating class to make a donation to Harvard. Although those "ambassadors" asking us to contribute acknowledged that many of us were in debt with tens of thousands of dollars of student loans, they insisted that the amount of money we donated was not important—it was the level of participation that mattered. The goal was to elicit contributions from as many of the graduating class as possible. A contribution of just five or ten dollars was valuable, we were all told, because it would signal our lifelong appreciation to Harvard for all that we had gained from being among the privileged few with Harvard degrees. We were told that our contributions would enable future students to attend Harvard and benefit from the same amazing experiences and opportunities we had. This narrative that we have a responsibility to "give back" and "pay it forward" to support future generations of Harvard students has been continually repeated ever since in the steady stream of fundraising communication that I receive from Harvard as an alum.

Given the devastating human suffering, ecological degradation, and economic injustice in the world, and given what I know about the role that elite universities play in the accumulation of wealth that exacerbates this suffering, degradation, and injustice, I have never donated money to either Harvard or Northeastern—or any institution of higher education. Instead, I have chosen to support organizations explicitly focused on climate justice and racial justice (and I donate regularly to the underfunded National Public Radio [NPR] in the United States). Given all the organizations and communities in desperate need of resources to support basic human needs, I am genuinely baffled by the idea that higher education institutions are so successful in convincing their faculty, staff, students, and alumni to donate to them.

According to *U.S. News and World Report*, a private media company that collects data to rank the hundreds of colleges and universities in the United States, higher education institutions report that on average 8% of their alumni donate, but among some private institutions more than 50% of their alumni contribute philanthropically (Moody 2020). The top ten for the two-year alumni giving rate include Princeton University, where 55% donate each year, and Dartmouth, where 44% of all alumni donate; these are both elite Ivy League institutions with long legacies of powerful and privileged alumni. Wellesley College, a small private women's college west of Boston, also has 44% of alumni contributing; and several small, remotely located, private colleges make up the rest of the top-ten list: Williams College (50%), Bowdoin College (47%), Amherst College (45%), and Carleton College (45%). Such high percentages of alumni giving suggest that fundraising efforts in many private US higher education institutions have been very effective in creating a culture of institutional loyalty. These institutions have established a social norm among their students of demonstrating loyalty and pride through consistent giving, and many alumni feel a responsibility and satisfaction from donating.

All of the top-ten institutions for alumni giving in the United States are residential campuses, which means that almost all the students live on campus in university-managed student housing throughout their years of studying. This residential college experience, a central part of privileged American culture, has become increasingly luxurious as higher education has been caught up in what is often characterized as an "amenities arms race" that has turned previously spartan student living into plush, luxurious, resort-like communities (Lieber 2021).

During these most formative years, when young adults are exploring their independence and struggling to define themselves in relation to the people and the world around them, many institutions of higher education are able to establish an institutional loyalty among their students that reaps direct financial benefits for years to come.

Lawrence Bacow, who was then the president of Harvard, expressed his public gratitude to a billionaire who donated $300 million in April 2023 by saying publicly that "generosity and loyalty" were two distinctive qualities of Harvard alumni. He also stated that he was "deeply and personally appreciative of the confidence he [the donor Kenneth Griffin] has placed in us—and our mission—to do good in the world." This statement by Bacow was surely intended to demonstrate confirmation and confidence that Harvard does "good," but this statement also reveals ambiguity, exposing an open question about whether and if Harvard does in fact "do good in the world." Is he suggesting that the $300 million gift confirms an otherwise unclear point? How do Harvard's current and future donors assess the "good" that Harvard does? Why does a billionaire decide to donate $300 million to Harvard, a university whose endowment is already over $50 billion?

Cultivating a Different Kind of Generosity

Critics have suggested that prestigious elite universities like Harvard are in fact creating more harm than good (Chung 2022) (see discussion on this in chapter 1). Calls for disruptive policies to diminish both the financial capital accumulation and the social capital associated with the wealthiest institutions in the United States are growing (Eaton 2022). Given the world's polarizing politics, inequitable economics, and worsening climate chaos, it seems prudent to explore how the perpetual, self-fulfilling resource accumulation at elite higher education institutions could be restructured for the common good. In the United States, political interest in taking action to disrupt the growing financial disparities among higher educational institutions is expanding on both sides of the political divide (Kim 2017).

What if ultra-wealthy universities like Harvard prioritized the cultivation within their community of a different kind of generosity? Rather than strategically encouraging students, alumni, and others to make financial contributions back to the university, what if they instead cultivated generosity based on the reciprocity that is central to so many indigenous and traditional cultures? Robin Wall Kimmerer, scholar, botanist, and best-selling author of *Braiding Sweetgrass: Indigenous Wisdom, Scientific Knowledge, and the Teachings of Plants*, calls for a cultural shift in society and in academic work toward restorative reciprocity, which involves nurturing an appreciation of nature's abundance of gifts that benefit us all (2013). Kimmerer urges us all to honor our individual and collective responsibility to respect and steward those gifts, reconceptualizing nature not as a resource for extraction but more as an elder relative whose wisdom we can learn from. Acknowledging the global crisis in mental health, Kimmerer suggests that gratitude can be a medicine for our ailing, lonely capitalistic society (Yeh 2020).

Cultivating compassion and gratitude for the world outside ourselves is an essential part of being human. Not only do we need to nurture our relationships with other people, especially those who are outside our immediate networks spatially and structurally (Smith 1999), we also need to connect with and appreciate nonhuman parts of the world, including animals, plants, trees, wind, sun, rocks, and stars (Kimmerer 2013). Financialization and the constant focus on economic and financial success defined by accumulating individual wealth has exacerbated the disconnection that so many human beings feel. Contemporary universities, and their competitive, individualistic cultures, are contributing to this disconnection, but a reimagined and restructured higher education system could be a countervailing force. Universities could be redesigned to intentionally dismantle the "othering" and dehumanizing that is disconnecting us through narrow competitive structures that encourage individual rather than collective wealth and power.

A core principle essential to understanding humanity and the earth's systems is the relationality of everything, the interconnectedness of all human beings as well as our links to the nonhuman world. Yet many higher education institutions devalue or ignore this inclusive relationality and instead focus narrowly on those relationships that might result in financial gains to the institution. Rather than teaching students to be loyal to their universities, what if higher education institutions focused more attention on demonstrating for students how to be loyal to the gifts provided from the nonhuman parts of earth's systems? Or what if universities demonstrated a commitment to the well-being of their local communities with the same sense of intense connection, pride, and loyalty that many are currently attempting to nurture among their students and alumni? And what if universities focused on demonstrating an institutional and community-wide commitment to ecological compassion and compassion for distant, disadvantaged people and vulnerable communities that might not be in their local neighborhoods?

This chapter expands on these ideas and encourages a reimagination of a regenerative and reparative financial structure for the higher education sector. To be able to contribute to the urgent societal need to redistribute and share knowledge, wealth, and power, a disruption in the current practices and policies of higher education institutions that concentrate and constrain knowledge, wealth, and power is urgently needed. Applying the "resist, reclaim, and restructure" framework adapted from the energy democracy movement, this chapter first focuses on resisting academic capitalism and the financialization of higher education. Then ideas for reclaiming higher education as a public good, rather than a private resource, are reviewed. The final section expands on specific possibilities for restructuring financial systems to support a reimagined higher education sector.

Resisting Academic Capitalism and Financialization

The strategic efforts to cultivate loyalty and a generous culture of philanthropy within universities emerges from, and reinforces, academic capitalism (Barry 2011) and the financialization of higher education (Eaton et al. 2016). As previously mentioned in chapter 2, academic capitalism refers to the commodification of knowledge, the commercialization of education, and the marketization of the higher education sector (Slaughter and Leslie 1999). Financialization of higher education refers to the ever-expanding role of finance and debt, including the expanding influence of financial markets, financial elites, and financial institutions in universities (Engelen, Fernandez, and Hendrikse 2014). Figure 5 represents the expansive implications of the financialized university, including growing university engagement in financial markets and real estate markets; the commodification of knowledge; a focus on revenue maximization; the weakening of faculty governance; overworked and overstretched academic, research, and support staff; increase in corporate and donor influence; students being treated as customers; and an explosion of both student debt and mental health issues.

The shift toward financialization of higher education has

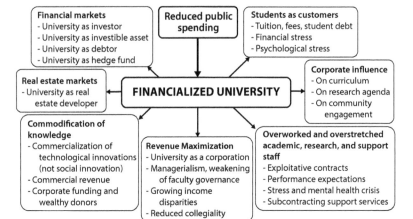

Figure 5 The Financialized University. Representation of the multiple complex implications of financialization in higher education.

emerged as part of the trend since the 1980s and 1990s to consider many contemporary societies as "knowledge-based economies," "learning economies," or "information societies" (Sokol 2003; Engelen, Fernandez, and Hendrikse 2014). Within this framing, higher education institutions were considered economically valuable sites of knowledge creation. When this way of thinking emerged in the 1980s, few recognized the degree to which higher education institutions themselves would be subject to market forces and financial markets. The surprising emergence of universities themselves behaving like businesses providing a paid service in which students are the customers was not explicitly envisioned nor was it predicted by many. When institutions need to compete within the higher education market to attract students, investments in attractive buildings and physical infrastructure can be viewed as more important than investments in academic programs and instructional staff.

The financialization of universities has been accelerated by the decrease in public funding for higher education (as mentioned in chapter 2). This decrease in public support is in response to government entities themselves being subject to financialization. To respond to market pressures, governments around the world have reduced public spending, including the financial support for higher education. In the United States, for example, state funding in 1980 accounted for, on average, 79% of public universities' revenue, while in 2019 that number had dropped to about 55% (Heller 2023). In the United Kingdom, public funding has also decreased, and the allocation of public funding is now linked directly to a research assessment exercise; this means that universities with more productive researchers get more funding. This has led to additional competition in recruiting and retaining highly productive researchers; some universities are "buying" key academics in a manner that resembles how sports teams compete over specific high-performance players. Although one of the original goals of the UK research assessment scheme, the Research Excellence Framework (REF), was to encourage higher-quality research across the board, the reality is that this performance-based research funding system has increased inequality and disparities within the UK higher education sector (Watermeyer and Derrick 2022). Less-resourced universities cannot compete with well-endowed institutions, and just as in so many other areas, the competitive landscape further advantages those institutions that are already advantaged. Beyond the United Kingdom, globalized financialization has dramatically increased inequities and disparities among higher education institutions within and among countries around the world (van Damme 2021).

Deep financial troubles for many UK universities were made worse by a 2015 policy change that abolished caps on how many students each institution could accept; this is resulting in overcrowding in some universities while others are closing their doors (O'Hara 2023). If universities are important institutions to the knowledge economy, why has this degradation of higher education institutions been allowed? Without a guarantee of sustained public funding, higher education institutions are financially vulnerable, and they look to other sources. This then opens the doors to corporate influence and other influential donors, including those representing fossil fuel interests.

The power of funders to influence academic work threatens the integrity and legitimacy of universities. Also, when so much time and effort within higher education is spent thinking about the university's budget and how to increase revenue for the institution, there is less time spent on strategic adaption of academic programs, research and learning approaches, and developing priorities that align with dynamic needs of society. In a competitive, market-based higher education landscape, the strategic work of university leaders prioritizes the financial success of the institution and its programs rather than the university's societal impact.

Many university leaders, including Lawrence Bacow, the former president of Harvard who lauded the extraordinary loyalty and generosity of Harvard alumni, perpetuate the idea within their communities that the financial success of the university is equated with positive social good for society. The internal logic of elite universities like Harvard is that the more money they have, the more good they can do in the world. This is the rationale for constantly cultivating donors—there is no amount of funding that is too much because the assumption is that the more wealth they accumulate the more good they will do. Given all the problems facing humanity that need fixing, there is infinite potential for Harvard to do good in the world.

Corporate influence, and the role of the fossil fuel industry in particular, on the research agendas of universities was discussed in chapter 4. Further to that, it is important to point out how corporate influence in the governance and management of higher education institutions has changed over the past thirty to forty years as university boards have become increasingly corporate (Lewis 2023; Cox 2013; Rowlands 2015). ExxonMobil, the US-based multinational oil and gas company, has an internal program to encourage its employees to donate and serve on the boards of the universities where they got their degrees. The vice chair of the Board of Trustees at Northeastern University, a former senior executive at ExxonMobil, shared (and is quoted on the Northeastern alumni page) that he was incentivized to contribute financially because his company had a "very, very generous matching gift program. . . . It's like throwing money away if you don't" (Northeastern 2023).

While academic capitalism is emerging differently in countries around the world, similar patterns are evident in many places. The competitive global market among higher education institutions has heightened attention to university rankings, comparative tools that provide public information and assessment of universities (Hamann and Ringel 2023). Although critique of higher education rankings has been growing, global and national rankings continue to impact research and teaching activities as well as student recruitment (Kaidesoja 2022).

An example of financialization includes the observation in a 2015 *New York Times* op-ed that Yale University had spent $480 million that year on fees for hedge fund managers to grow the university's already massive endowment—while the university only spent $170 million on tuition assistance and fellowships for its students. This example demonstrates how fund managers rather than students benefit from these large endowments. Despite much critique and widespread concern about the risks and negative impacts of financialization of higher education (Russel, Smith, and Sloan 2016; BER 2022; Foroohar 2016), policies and practices have been reinforcing, rather than reversing, the trends of financialization (McGeown and Barry 2023). Many education policy advocates have been calling for massive public investments in higher education, making the case that universities are, overall, an asset to society rather than a cost on the balance sheet (Foroohar 2016).

The stark economic inequities that are steadily getting worse in capitalist, free-market societies are also represented with worsening internal economic disparities within academic institutions. In most colleges and universities in the United States, a highly paid president and a growing team of multiple well-compensated senior administrators lead the university, while the educational programs rely on an academic staff increasingly made up of faculty and staff on precarious short-term contracts and low pay. Food service, custodial, and maintenance staff, who are critical to the operations of higher education institutions, are often paid minimum wage. Increasingly, these positions are contracted out to external companies, so these workers are not officially university employees and are therefore ineligible for many of the benefits afforded to other university workers. Constant striving for growth in the research enterprise at most universities results from the financial rewards to the institution when researchers secure external research funding; because each external grant comes with a high "overhead rate," which is over 50% in some universities, a high proportion of external funding does not go directly to the researcher but goes to support the university's basic operating costs.

The increasing cost of attending university is another consequence (and demonstration) of the financialization of higher education. Although the Nordic countries in Europe provide publicly funded tuition-free third-level education—and Norway, Finland, and Denmark also provide payments to support students' living expenses while they are enrolled in university (Valimaa 2015)—in most places in the world, the cost of attending university is the biggest barrier to access. In the United States, rapidly increasing costs have been enabled by a federal student loan subsidy program that encourages students to go deeply into debt to cover their costs (Robinson 2017). At Northeastern University, the higher education institution where I used to teach in the United States, annual tuition was about $62,000 for the 2023–2024 academic year. This high tuition rate means that a majority of students have to take out tens of thousands or even hundreds of thousands of dollars of loans to cover the cost of their university education.

Rapidly growing high compensation packages for university presidents is another consequence and demonstration of financialization of higher education. In the United States, high salaries result from competition for charismatic, sought-after academic leaders and more corporate executives on the boards. Salaries of the highest paid university presidents in the United States are published each year by the *Chronicle of Higher Education*. The highest salaries have been frequently between $1.5 and $3 million per year (*Chronicle of Higher Education* 2022), although the total compensation may be higher. Beyond the base salary, total compensation can include bonuses, housing, and various other expenses and in some cases even a chauffeur or a cook.

University staff throughout the UK system were striking with some regularity in 2022 and 2023 to advocate for better pay, pensions, and working conditions. In the United States, faculty unionization exists primarily in public institutions because the law regarding the right to unionize in private colleges and universities is uncertain; faculty in private universities are considered "managerial employees," so they are explicitly excluded from coverage under the National Labor Relations Act (AAUP 2023). Internally within universities, the widening disparities in compensation between the senior leadership and a growing number of precarious, low-wage academic staff is a sensitive and highly contentious issue.

The wide discrepancy in economic realities within the Northeastern University community was acknowledged explicitly during the early days of the pandemic in 2020 when the university president reported to the university community that he would contribute 20% of his annual salary to a new COVID-19 fund created to support students facing economic hardship because of the pandemic and to support research related to studying the crisis. At this time, it was also announced that each of the university's senior vice presidents and academic deans would also contribute 10% of their salaries to the same fund. This public recognition of the economic inequities within the university, demonstrated by the one-time gestures toward redistribution within the community, was a rare occurrence inspired by a moment of crisis. Day to day, the juxtaposition of the extreme wealth and high incomes of university leaders with the economic precarity of many students and staff is ignored, and community conversations about these disparities are often discouraged.

Resisting academic capitalism and the financialization of higher education includes creating a culture of transparency and accountability and welcoming conversations about financial issues within the community. Resistance also comes from efforts to reveal how the constant focus on revenue, growth, and expansion undermines academic initiatives and constrains the kinds of impact possible (McGeown and Barry 2023). Efforts to expose and minimize corporate influences, including the fossil fuel divestment movement (Healy and Debski 2017; Mikkelson et al. 2021) and other efforts to reveal the strategic investments by the fossil fuel industry designed to leverage the power of universities to legitimize sustained fossil fuel reliance around the world (Banks 2023), are also important actions for resisting academic capitalism.

Because of the power, wealth, and associated intimidation of the fossil fuel industry, it is difficult for higher education institutions to resist these influences. One example of a courageous organization coordinating the collective effort to resist fossil fuel interests in universities in the US context is UnKoch My Campus, a nonprofit organization whose mission is to preserve democracy by protecting higher education from actors whose expressed intent is to place private interests over the common good (UnKoch My Campus 2020). UnKoch My Campus was founded in 2013 in response to the realization that donations from the Koch family were influencing the curriculum, research agendas, and the hiring and firing of faculty on multiple campuses across the United States. Students at George Mason University, Florida State University, and University of Kansas organized campus protests, and when the university administrations refused to reveal the details of these donations, the students recognized the need for coordinated resistance. Since its founding, UnKoch My Campus has expanded to become a national nonprofit organization providing resources to other campuses around the country and internationally. The organization investigates and audits relationships between wealthy donors, corporations, and educational institutions to reveal the strategic investment promoting private interests over the common good and provides training on how to resist. Recognizing that public opinion and public policy are shaped by the pursuit and production of knowledge in higher education institutions, UnKoch My Campus is leading the way in resisting corporate influences throughout society (Banks 2023).

An additional form of resistance includes university commitments not to do business or affiliate with certain companies. For example, Brown University has committed to avoid and decline all financial interactions with organizations that promote misinformation (Brown University 2022). With this commitment, the university has acknowledged that conducting business with individuals or organizations that directly support the creation and dissemination of science disinformation—that is, knowingly spreading false information with the intent to deceive or mislead—is contrary to the university's mission of advancing knowledge and understanding (Brown University 2022). Brown University is explicitly acknowledging, with this commitment, that universities are vulnerable to being manipulated if they are not careful and deliberate in determining what organizations and individuals they want to support through their financial transactions.

The protests and widespread campaigns on many US and European Union campuses demanding that universities disclose and divest from companies profiting from the Israeli government's militarized devastation in Gaza provide another example of efforts to resist the financialization of higher education (Stephens 2024). As the violent destruction of current extractive and exploitative systems get worse, calling for divestment offers students a way to take action to advance transformative change toward a more just and equitable future. As more students recognize the misalignment between their universities' claims to be addressing the grand challenges facing humanity and the lack of engagement with structural change on their campuses, divestment campaigns are likely to continue to expand.

Reclaiming Financial Flows for the Collective Good

While many higher education leaders are increasingly focused on the financial health of their institutions, the crises of the world are creating new opportunities to reclaim a financial model in higher education that allows universities to become central nodes for redistributing—rather than concentrating—wealth. Higher education institutions could collectively accept the extensive social science research that shows that growing economic inequities are the cause of deteriorating conditions for humanity (Malleson 2023; Kenner 2019; Hernando and Mitchell 2023; Piketty 2015). If the mission of academia was to reduce, rather than reinforce, economic inequities, higher education could be slowing down and reversing, rather than worsening, the intersecting crises facing humanity (McGeown and Barry 2023).

Imagine if promoting economic justice was a goal of universities. Rather than catering to wealthy donors, corporate interests, and students from wealthy families (Eaton 2022) the higher education sector could redefine its public good mission to be a central resource for achieving society-wide economic equity and an orchestrator of the redistribution of wealth throughout society. Economists, bankers, and business/management scholars could shift their attention to provide analysis and recommendations on transformative innovations of monetary and fiscal policies that would curtail the continued existence of billionaires, ensure everyone had access to housing, food, and education, restrict corporate profits, and develop wealth and income maximums and minimum for both individuals and organizations. Higher education could become a hub for monetary policy innovation exploring the range of policies that central banks could deploy to incentivize financial markets and the banking sector to advance economic equity and climate justice (Sokol and Stephens 2022; Stephens and Sokol 2023). Rather than reinforcing the current financialized power dynamics that have been constraining research on financial innovations for the public good, academic institutions could be co-creators of financial research for climate justice with a goal of developing and implementing innovative financial tools for economic justice, climate stability, and investments for the common good (Positive Money Europe 2020). Instead of partnering with big corporate banks, universities could partner with community-based credit unions and organizations like Positive Money, a nonprofit organization in Europe, the United Kingdom, and the United States, that conducts research and advocacy on reimagining money, banks, and the economy for the well-being of people, communities, and our planet (Positive Money 2018).

The range of different fiscal policies to be explored is broad, including tax increases on extreme wealth and expanded government spending and public investments on everything required for a healthy society. Ensuring access for all to free public education at every level, including higher education, has to be prioritized in any effort to reduce inequities by redistributing wealth and power. In societies throughout human history, the social value of providing free education has been recognized, with primary and secondary education being supported first in the nineteenth and twentieth centuries, and now in the twenty-first century many countries around the world do provide free third-level education. Just as most countries already ensure that providing free public elementary and free public secondary schools is essential for the economic and political health of the society, providing free public higher education, or third-level education, will strengthen future prospects for any society (Harris and Mills 2021).

As the global movement for free university education grows, emerging innovative institutions are circumventing the slow pace of change in mainstream universities. For example, University of the People (UoPeople) is a disruptive innovative institution that offers 100% online, tuition-free degree programs that are modular and flexible to be inclusive and accessible to students around the world. UoPeople, which was founded in 2009 and accredited in the United States in 2014, is based on the premise that access to higher education is a basic right that promotes peace and global economic development. In 2023, more than 126,000 students from more than two hundred countries, including 16,500 refugees, were enrolled. While there is no cost for tuition, the university does have a one-time application fee of US$60, an assessment fee per course completed of $120 for undergraduate courses and $300 for graduate programs, and scholarships available for those who cannot afford those fees. By removing most of the cost to attend university, UoPeople is opening up alternative options within the higher education sector.

Other innovative approaches to reclaiming the financial flows of higher education include the idea that private universities could be reclaimed by the state and restructured as public institutions (Chung 2022). This suggestion goes against current trends in the United States, where some are calling for more public universities to be privatized (Riley and Piereson 2020). The extreme accumulation of wealth among the most well-endowed universities, however, is becoming increasingly obscene and difficult to justify. The creation of some kind of institutional wealth cap, just as some are proposing wealth caps for individuals, could provide a mechanism for disrupting the way that elite universities are currently concentrating wealth and power.

In the United States, both private and public colleges and

universities are classified as not-for-profit organizations, so they are exempt from paying taxes. Unlike any other property developers in a city or town, a university can buy and develop real estate without having to pay tax on the value of the property. This is a large financial benefit to universities, and conversely this creates a huge financial burden to the local municipality because cities and towns lose critical tax revenue as universities expand within their jurisdictions. Given the reliance on property tax to provide local municipal services for communities, the fact that universities do not have to pay taxes on their properties means that they are often a drain on local city budgets (Baldwin 2021). One specific mechanism for universities to contribute to redistributing, rather than concentrating, wealth would be if municipalities were able to reclaim the tax revenue that they do not receive from the properties owned by higher education institutions.

Part of reclaiming the financial flows of universities also includes reprioritizing the work environment for faculty and staff. In universities throughout the world, faculty and staff are reclaiming their voices and standing up for their own wellbeing in multiple ways (Urai and Kelly 2023). Resistance to the growing pressure of expanding expectations and requests to do more for the university with less resources is taking shape in different initiatives, including strikes and protests, resignations, and more subtle forms of rebellion. Recognizing that the pressures put on faculty and staff in the current competitive financialized universities is creating a mental health crisis within higher education, more universities are offering mental health resources. Despite the widespread recognition of the increasingly stressful working conditions, very few universities are implementing structural or financial changes to reduce the growing precariousness of academic positions.

To unleash the potential for universities to catalyze and support economic justice, redistribution of wealth, and a climate stable future for all, reclaiming the financial independence of higher education needs to become a collective goal. Higher education institutions are not only objects or victims of financialization, many are also active participants reinforcing financialization throughout society. For universities to be able to serve the common good and remain independent from manipulation and corporate capture, their role in financial markets, in real estate markets, and in the commodification and commercialization of knowledge needs to be minimized, and public funding for universities needs to be increased.

To illustrate this point, consider that some universities invest some of their money in fossil fuel–related stocks. When they do this, universities are literally investing in the success of the fossil fuel industry, which is counter to the science of climate change. This is why fossil fuel divestment campaigns and fossil fuel–free advocacy among students and faculty of universities has been so strong (Stephens, Frumhoff, and Yona 2018). Although thousands of universities around the world have divested from fossil fuels (Mikkelson et al. 2021), the fossil fuel industry continues to strategically invest in higher education to leverage the power of universities to legitimize and conceal the ecological devastation of fossil fuels.

Restructuring for Wealth Distribution

Accepting that the current financial realities shaping higher education are constraining the societal impact of universities and also concentrating, rather than distributing, wealth, this section reviews specific ideas on alternative financial structures. These ideas include internal restructuring within higher education institutions as well as external restructuring of public investments to fund higher education. Each of these ideas relies on, and has implications for, politics, policy, and economics both within and outside academic institutions.

The most fundamental principle for restructuring funding for higher education for wealth distribution is to ensure strong, reliable public funding. Ensuring public funding is essential for universities to be independent of financial interests and contribute to supporting wealth distribution efforts and community wealth building. Community wealth building involves aligning policies, practices, and institutional commitments to strengthen communities by building local wealth through economic inclusion, broad-based community ownership, and control of assets with a goal of ensuring well-being of all residents (Lacey-Barnacle, Smith, and Foxon 2023; McMahon 2020). Local ownership of affordable housing, cooperatives, employee-ownership models, and land trusts are prioritized to retain both wealth and decision-making within the community. Creating local jobs that prioritize worker well-being and fair wages, benefits, and career pathways is also prioritized along with local procurement and renewable and regenerative production and consumption of food, energy, and other material goods. Research on community wealth building has highlighted the critical role for anchor institutions, including universities (Lacey-Barnacle, Smith, and Foxon 2023).

If universities are reimagined as critical infrastructure focused on supporting community wealth building and providing region-specific support and responsiveness to communities, then strong public investments in geographically distributed university systems could provide essential capacity building for climate resilience, revitalize local empowerment, and drastically reduce other costs and expenses for communities. In this way, increasing and expanding public funding for higher education can be considered both a critical climate policy priority and an important investment tool to reduce climate vulnerabilities and alleviate economic precarity in communities throughout the world.

Given the expansive benefits of providing strong public funding for higher education during this dynamic time of urgent need for community capacity building to confront intersecting crises, public investments in universities would pay off quickly and have a high return on investment. If universities are restructured to orient their missions more directly with community partners in co-creation and collaboration, the societal expectations for how, when, and in what ways universities are directly serving the needs of communities can be much more ambitious than they are currently. The next chapter (chapter 6) explores in more depth the possibilities centering communities and distributing and decentering universities within and among local and regional communities. One outcome of increased public funding would be an elimination of the need for universities to be constantly fundraising. This would result in more concentrated attention and time spent co-designing, co-defining, and collaboratively implementing dynamic priority areas that center communities.

Another proposed idea for restructuring higher education is to municipalize universities, which refers to the process of transferring the control and administration of higher education from national levels of government to the municipal or local level. Advantages of municipalizing higher education could include greater local control and responsiveness to the needs of the community and increased community engagement and participation in decision-making. It would structurally embed the university in the city, town, or region where it is located, fostering close collaboration and co-design and co-production of knowledge to meet community needs. This idea is quite radical, and of course there are many potential challenges with municipally controlled universities, including the management and coordination of the level of funding, standards, accreditation, and recruitment. But if universities were to become critical infrastructure for every town and city, a new municipal approach with local control and management could be an effective structure for universities to become more locally oriented and engaged. Around the world, there is a growing movement toward municipalism and democratic public ownership of other collective assets and infrastructures (Sareen and Waagsaether 2022; Wenderlich 2021); and radical municipalism has been recognized as a strategy in times of crisis (Roth, Russell, and Thompson 2023).

Within countries like the United States where there are multiple private universities in addition to the public state-managed universities, another strategy could be to move all higher education institutions into the public sector—that is, abolish private universities. If a new mission for universities included facilitating community wealth building, the private university model would be outdated and contrary to the goals of the institutions. Converting private universities into state-managed and publicly funded public higher education systems may seem outrageous to those who are loyal to, and benefited directly from, existing private universities, but this idea has been suggested (Chung 2022). Putting this option on the table catalyzes new ways of assessing and imagining universities as collective, social justice organizations structured to prioritize community care and community well-being in an era of cascading intersecting crises.

Another provocative proposal for restructuring higher education for wealth redistribution is to create a wealth cap for individual universities, as mentioned earlier. Here I am suggesting that an institution like Harvard, for example, an institution that has an endowment of over $50 billion, would no longer be permitted to accumulate that much wealth. Imagine, for example, that there was an institutional wealth accumulation cap of $1 billion; then the remaining billions could be distributed to fund a global network of regional higher education institutions that focus on local needs in every region of the world. This idea of investing in regionally and spatially distributed universities is also synergistic with the goal of universities facilitating strong local relationships and networks of relationships (discussed more in chapter 6). One result of universities being community centered in their local regions is that students would establish strong local connections and relationships in the communities in which the universities are located. Those students may have additional incentives to stay in that community as a long-term place to live. Distributing universities in less-populated rural areas could, therefore, also serve to distribute the general population in areas outside densely populated cities, where many universities are currently located. Leveraging local connections and strengthening place-based relationships provides important learning opportunities, particularly in relational knowledge, a specific kind of knowledge that has been historically neglected and dismissed within many contemporary universities.

An additional proposal for restructuring the finances of higher education in the United States would be to renormalize pride in paying taxes to contribute to the collective good. What if people enjoyed paying taxes because they knew that their tax contributions were supporting the public universities that are essential for sustaining a caring and compassionate society? This requires a cultural shift away from the current dominant narrative that fiscally responsible Americans do whatever they can to minimize how much tax they pay to the government. Because universities are not-for-profit organizations, donations to universities are tax deductible. The many

Americans who go to great lengths to minimize paying taxes often make tax-deductible charitable donations to universities to reduce their tax bill (Stewart 2023). Instead of wealthy Americans donating money to elite private universities, what if we restructured the tax system so that those same wealthy Americans were incentivized to instead contribute their money to public universities. The current system in the United States concentrates wealth among the wealthiest universities because individuals are rewarded for making multimillion-dollar donations to already privileged institutions serving already privileged students.

A related restructuring approach involves changing tax law so that billionaires are heavily taxed. This would constrain and cap a maximum amount of accumulated wealth that any one individual can hoard. Multiple economists who conduct research on alternatives to the perpetual growth model of capitalism have proposed restructuring fiscal policy to ensure that there is both a maximum level of income and wealth for those with lots of resources and a minimum level of income and wealth provided to everyone so nobody is left behind (Raworth 2017; Hickel 2023).

These restructuring ideas, including municipalizing universities, distributing universities, and ensuring full public funding for universities, would also serve to dismantle the competitive marketplace for higher education and disrupt the higher education ranking systems. Although university ranking systems are often criticized as being superficial and fraught with bias, most universities and many prospective students look to the rankings for legitimate comparative assessment of higher education institutions. The contentious landscape of rankings in higher education reflects the superficial realities of academic capitalism (Hamann and Ringel 2023). In response to criticisms about college rankings, the *New York Times* developed a novel independent and interactive college ranking system that allows potential students to define for themselves what metrics matter most to them (Bruni 2023). In 2023, Utrecht University in the Netherlands got international attention for opting out of the Times Higher Education World University Rankings list, citing concerns about the emphasis on scoring and competition. This was significant particularly because Utrecht University had previously ranked in the top one hundred universities.

Another restructuring approach to consider how universities could contribute to distributing rather than concentrating wealth is a conceptual restructuring. What if we collectively redefined wealth not as the hoarding and accumulation of money, which leads to excessive consumption, loneliness, waste, violence, and disparity, but as regenerative, embodied wealth associated with our health, the social capital that we have in our networks of relationships, local knowledge, and our closeness and appreciation for nature (Jain and Senggupta 2023)? This alternative definition of wealth prioritizes the regenerative potential of life and humanity. What if instead of focusing on higher education as a path to a steady livelihood defined by earning money in a steady job, higher education focused on nurturing *alivehoods*, defined as finding daily work that replenishes and nurtures various forms of real wealth, including health, social capital, nature, and local knowledge? Alivehoods refers to work that goes beyond measuring success and impact based on the typical financial pay packages, work that takes us beyond our fears and scarcity by revaluing and rebuilding a collective field of trust, dignity, and a sense of inner fulfillment (Jain and Senggupta 2023). If universities were able to nurture this sense of creative abundance, then students and communities could start to imagine and make different choices for themselves and the planet. By expanding learning of knowledge and wisdom of indigenous cultures, universities can be places to support a paradigm shift in honoring a very different definition of wealth. Robin Wall Kimmerer (2013) reminds us that "wealth among traditional people is measured by having enough to give away. Hoarding the gift, we become constipated with wealth, bloated with possessions, too heavy to join the dance."

The Cooperative, Worker-Owned University

Recognizing how knowledge and action—and theory and practice—are intricately linked and self-reinforcing, the praxis of universities, including their governance and financial structures, informs, constrains, and determines universities' potential impacts on students, staff, and communities. To implement a reimagined new and different role for universities requires, therefore, a reimagined new and different structure for universities. One specific structure with historical and contemporary relevance to universities is the cooperative, worker-owned university. Cooperative, worker-owned organizations are democratically controlled by their employees; the workers collectively own and manage the organization, sharing decision-making, responsibility, control, profits, and benefits (Wiksell 2020; Baskaran 2015).

I have been personally inspired by the power of cooperatives through my experiences living in a rural community in Donegal in the Northwest of Ireland, where, since 1906, the town of Dungloe has had a cooperative society, The Cope. A 1939 autobiography by one of its founding members and community leaders, Patrick Gallagher, known as Paddy the Cope, reflected on the societal potential of cooperatives:

> If the business of the world was run by co-operative societies there would not be the terrible slaughter that is going on in the world. . . . If all were organized in cooperative societies there

would be no motive for wars as the co-operative motive would be to help all the people of the world. . . . Can anyone imagine what such a body could do for humanity? Think of all the scientists in the world sitting at a co-operative conference studying what is best for the human race, instead of acting today in their individual capacity in the interests of individuals investing weapons to destroy the other fellow. (348)

Applying the cooperative model to higher education provides an opportunity to consider a specific alternative structure for universities that could support and facilitate a transformative climate justice orientation in ways that academic capitalism is unable. The Mondragon University (Mondragon Unibertsitatea) in the Basque Country of Spain provides an example of a cooperative university. This university, founded in 1997, has a commitment to social transformation and is part of the Mondragon Corporation, a network of cooperatives based on democratic decision-making and putting the rights and well-being of workers first (Mondragon Unibertsitatea 2023). Mondragon University is structured for social accessibility and combining work and study, and the university is committed to the "environment, our society and our time" (Mondragon Unibertsitatea 2023). The educational model is based on cooperative values, principles of cooperative entrepreneurship and innovation, humanism, and solidarity. As of 2023, the university has four faculties and well over six thousand students (five thousand plus undergraduate and one thousand plus postgraduate).

Deeply rooted in the Basque Country, Mondragon University is committed to the Basque language and culture while, at the same time, being open to the changes demanded by the twenty-first-century society. Mondragon University is also making important contributions to renewable energy efforts:

in 2023 it inaugurated a new medium-voltage laboratory aimed at improving the wind energy sector (Mondragon Unibertsitatea 2023). One important feature of all Mondragon cooperatives is the egalitarian wage structure in which top management does not get paid more than six times the lowest-paid worker (Kelly and Massena 2009). This aspect—a simple principle defining the wage structure of the institution—could be implemented in any university system even without the cooperative, worker-owned model. With growing attention and research on income and wealth caps as a key mechanism for addressing worsening economic inequities (François, Mertens de Wilmars, and Maréchal 2023), exploration and experimentation of these ideas within the higher education sector are likely.

Given the legacy of universities being owned and managed by faculty, the idea of a cooperative, worker-owned university is not actually that far from existing or historical models. Trinity College Dublin, for example, the oldest university in Ireland, was founded in 1592 by Queen Elizabeth I as part of England's colonial project. Unlike the other contemporary universities in Ireland that are public, Trinity College has a unique legal and governance structure: it is an autonomous corporation with a charitable status consisting of the provost, the fellows, and the scholars. This unusual structure is still reflected in the long legal name of the university, which reads, "The Provost, Fellows, Foundation Scholars and the other members of Board, of the College of the Holy and Undivided Trinity of Queen Elizabeth near Dublin."

Unlike in many other universities, the provost, who is the head of the college (the equivalent of the university president in the US system), is *elected* by the university community for a ten-year term. Members of academic staff are elected as fellows based on their research achievements. The board, the main decision-making body of the university, includes representatives of all main constituent bodies of the college community—namely, fellows; non-fellow academic staff; professional, technical, administrative, and support staff; and student union members. Indeed, one of the explicitly recognized governance principles of the university is collegiality (which manifests itself in the participation in its governance by members of the Trinity community). The board reaffirmed the value of collegiality and plurality in 2020, recognizing that these values ensure that the range of experiences and perspectives of the community enhances the quality of institutional decision-making (Trinity College Dublin 2023).

While not a worker-owned cooperative, Trinity's governance model still differs from the corporatized, top-down models of governance that have become widespread in other higher education institutions. This democratic governance structure may have been responsible for Trinity College Dublin being the first of the historic universities of Britain and Ireland to admit women in 1904 (Parkes 2004); the forty-fifth provost of Trinity, Linda Doyle, was the first woman elected to the position in 2021.

Funding and Finance for a Culture of Care

Northeastern University's Giving Day takes an immense amount of time, creativity, and planning; it is a major initiative coordinated by the Office of University Advancement, the fundraising team that includes a staff of approximately two hundred professional fundraisers. But imagine if rather than rallying the university community to fundraise for itself, all that time and effort was directed toward developing a culture of care? What if all the Giving Day ambassadors were asked to send community-wide emails asking their colleagues and peers to check in with a neighbor or a neighboring community instead of contributing cash? Rather than asking the Northeastern community to donate money to advance internal Northeastern activities, what if Giving Day was a designated day for the Northeastern community to "give" to a cause or an organization or a group outside of the Northeastern community? The inward self-referential, financial focus in so many US universities detracts from the intellectual, cultural, and social justice potential of these institutions. The financialization of every aspect of academic life in these universities creates an institutional environment where communicating success is prioritized, and critical reflection on any negative impacts of university activities is discouraged.

In the book *A Beginner's Guide to Building Better Worlds*, a collective team of diverse first-generation authors encourages us all to see ourselves as co-creators and shapers of the societies, cultures, communities, and institutions that we are a part of (Gahman et al. 2022). Drawing from the resistance and mutual aid experiences in Zapatista territory, they explain that neoliberalism has us believe that individualism, competition, private property, and profit-seeking are the natural state of things, and that markets will continue to dictate our work, our productivity, nature, and time because these things can all be commodified and measured in monetary terms (Gahman et al. 2022). But there are so many other ways of conceptualizing humanity and our relationship with the earth and its planetary boundaries, and universities have a critical role to play in expanding the plurality of imagined futures.

Universities are essential for legitimizing and elevating the experiences and perspectives of people and communities whose understanding of the world is more expansive than the dominant neoliberal model. Neoliberal thinkers tend to dismiss and omit the structural forces causing disparities and inequities, and they focus instead on prioritizing personal responsibility; neoliberalism, therefore, places the blame for poverty, sickness, and deprivation on individuals rather than on the societal systems (Gahman et al. 2022). Contempt for, and disdain toward, the poor is fundamental to neoliberal thinking, and neoliberal universities are perpetuating this dangerous worldview. Neoliberal economics and neoliberal thinking reinforce wealth supremacy, the term Marjorie Kelly has coined to describe the pervasive bias in policies, practices, and priorities for those who are wealthy (Kelly 2023). Just as university practices and university research have reinforced white supremacy, university practices and university research are reinforcing wealth supremacy.

In contemporary capitalistic societies, higher education institutions have been leveraged as key nodes in the complex interconnections among knowledge, wealth, and power (Sokol 2003). For universities to play a larger role in disrupting the destructive path that humanity is on, the vicious cycle in which higher education continues to concentrate knowledge, wealth, and power needs to be broken (figure 3). Instead, the potential for higher education to distribute and regenerate knowledge, wealth, and power can be realized by restructuring higher education funding and finance to promote a culture of collective care.

Both research and practical experience confirm that organizations struggling for economic survival are more constrained than those flush with cash (Hurth and Stewart 2022). When university leaders are narrowly focused on the central issue of economic survival, they are unable to engage in inquiry and exploration of larger societal transformations (Hurth and Stewart 2022). In this era of intersecting crises, expanding human suffering, and ecological devastation, sustained, large public investments in a distributed higher education system would be a powerful force toward creating a brighter future for all. Imagining publicly funded higher education systems designed to promote a culture of collective care with a commitment to advancing both community and ecological health is a key part of the transformative vision of climate justice universities.

CHAPTER 6

Local Empowerment and Global Solidarity

The vision of climate justice universities includes higher education systems that are committed to serving the needs of local and global communities. A reimagined mission of higher education could be to empower people locally while connecting in solidarity with other communities throughout the world. This mission and vision decenters the university itself and recenters community needs. While previous chapters discussed how climate justice principles intersect with teaching (chapter 3), research (chapter 4), and finance (chapter 5), this chapter highlights the need for a radical transformation of what is usually referred to as university "outreach" or "engagement."

When Hurricane Maria, a devastating Category 4 storm, made landfall in Puerto Rico on September 20, 2017, the campuses of the University of Puerto Rico were severely damaged, with buildings flooded, roofs blown off, and energy systems destroyed. In the immediate aftermath of one of the deadliest disasters in Puerto Rico's history, the university was also a first responder, playing a crucial role in providing humanitarian support. Its campuses served as centers for distributing food, water, medical supplies, and other essential resources to local communities. Without power and water for several months, resuming academic activities was difficult, and many students, staff, and faculty had to relocate. In response to the devastation, a network of academic colleagues from mainland US universities traveled to Puerto Rico with a goal of offering assistance and resources in many forms, including quick-response research support, service learning, student exchange, humanitarian relief, relocation assistance, and financial support. In the crisis conditions, the coordination of leveraging and distributing the different offers of support in a timely and effective way proved challenging. Despite the well-intentioned efforts, the post-Maria experience revealed weaknesses in several aspects of university-community relationships. In response, colleagues at the University of Puerto Rico–Mayagüez started the RISE Network, an inter-university collaborative convergence network determined to improve and coordinate university interventions in pre- and post-disaster environments.

RISE is a knowledge-sharing network focused on developing a new architecture of relationships between universities and communities for the collaborative enhancement of community resilience (RISE 2023). With institutional and individual members throughout the United States, RISE facilitates formal and informal connections among universities and communities through convergence dialogues, convenings, and workshops (figure 6). The largest in-person RISE convening, hosted at State University of New York (SUNY) Albany for three days in November 2019, included close to four hundred researchers and educators from more than 114 institutions to collectively explore how higher education institutions can collaborate with one another and with communities to strengthen preparedness, response, and recovery in the face of growing threats posed by climate disruptions and extreme weather.

Since the devastation in Puerto Rico after Hurricane Maria, the cofounders of RISE, environmental sociologist Marla Perez-Lugo and public administration scholar Cecilio Ortiz-Garcia, have moved from the University of Puerto Rico. Following several years of temporary and precarious positions, they landed at the University of Texas at Rio Grande Valley (UTRGV) in

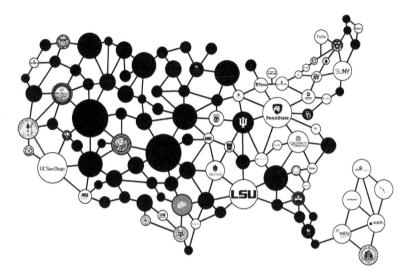

Figure 6 RISE Network Map. A representation of the network of universities contributing to the RISE Network, including Arizona State University, Bates College, City University of New York, Clemson University, Duke University, Harvard University, Indiana University, Louisiana State University, Loyola University, New Jersey City University, Northeastern University, Pennsylvania State University, Sonoma State University, State University of New York, Texas A&M University, Tufts University, University of Arkansas, UC San Diego, University of Central Florida, University of Colorado, University of Delaware, University of Maryland–Baltimore County, University of Pennsylvania, University of Pittsburgh, University of Puerto Rico, University of Texas at El Paso, University of Texas at Rio Grande Valley, University of Virginia, University of Washington, Utah Valley University, Virginia Tech, and Willamette University. *Source:* RISE Network, https://therisenetwork.org/index.php/organizational-structure/.

Edinburg, Texas, one of the largest Hispanic-serving institutions of higher education in the United States. Given our shared interests in understanding how disruptive events can be leveraged for climate justice transformation, I have collaborated with this inspiring academic couple as part of a larger research team that includes several other colleagues since 2018 (Kuhl,

Stephens, et al. 2024; Perez-Lugo, Ortiz-Garcia, and Valdes 2021), and I serve on the RISE Network advisory council. I had the privilege of visiting UTRGV and participating in a RISE convening in March 2022, an event that the organizers called "a transboundary convergence dialogue," focused on exploring the "power failures" of the climate disruption in Texas in 2021 that caused a deadly energy blackout through the lens of energy justice and energy democracy. By centering community voices and prioritizing the devasting health vulnerabilities associated with power failures during extreme weather, this event decentered university expertise and provided a community-centered forum. Climate disruptions in Texas are resulting in both more frequent extreme cold weather that the energy system is not designed for and longer and more intense blistering heat waves that increasingly push the Texas power grid to its limits (Tigue 2023).

In summer 2023, Marla and Cecilio established a new hub for the RISE Network at UTRGV by creating the Center for Community Resilience Research Innovation and Advocacy (CCRRIA), focused on innovation on university-community relationships not from the perspective of the university but, more importantly, from the perspective of vulnerable communities. This decentering of the university itself, and the recentering of the needs of vulnerable communities, is a paradigm shift that is desperately needed in universities around the world. The RISE Network and UTRGV's new center are pioneers, leading the way in the higher education paradigm shift, centering the needs of local and regional communities.

In addition to devastating climate disasters forcing universities to reconsider their role in their local communities, growing economic precarity in the cities, towns, and regions where universities are located creates another imperative for higher education institutions to reimagine and restructure their partnerships and interactions with local communities. In response to worsening poverty and the associated health and safety disruptions of communities in crisis, many universities are building higher walls or stronger fences and increasing their security measures. But relying on a fortress approach reinforces the coloniality of "othering" those who are not officially part of the university community. Until and unless universities shift their mindset toward centering the needs of local communities and co-creating alternative futures through local empowerment, universities run the risk of becoming increasingly unappealing and uncomfortable places characterized by cognitive dissonance that perpetuates an out-of-touch, isolated narrow mindset.

To leverage the transformative potential of higher education institutions, universities need to collaborate more extensively and co-create more directly with nonacademic partners and communities. While expanding new trusted relationships with communities and local organizations, as part of a recommitment to higher education's public mission (Papadimitriou and Boboc 2021), universities need to simultaneously decline partnerships and disassociate from corporate interests and others who are intent on sustaining the status quo (Oreskes 2015; Stephens, Frumhoff, and Yona 2018; Westervelt 2023). This chapter explores new ways of thinking about university-community interactions for transformation with a focus on the power of a justice-centering relationships framework (Quan 2023). A justice-centering relationship framework acknowledges the dominant power structures that reinforce an assumed imbalance of knowledge and wealth within university-community interactions; this imbalance tends to prioritize benefits of partnership to the university rather than benefits to communities. Resisting this hegemonic pattern requires developing long-term, genuine, movement-building, place-based, community-centered institutional strategies that transcend individuals and focus on community dignity, community trust, and community needs rather than university outcomes (hooks 2003; Quan 2023).

As we reimagine how universities could commit to distributing rather than concentrating knowledge, wealth, and power, new ways of empowering and co-creating with local communities are central to the transformative shift that is needed. Expanding from the example of the RISE Network, this chapter introduces other ongoing initiatives that are redefining the architecture of university-community interactions. While many universities are currently making efforts toward expanding community engagement, the ideas discussed here expand on mainstream resources to encourage academics to consider their individual engagement beyond the campus (Beyond the Academy 2022). Similarly, recognizing that many individual students, staff, and faculty sustain empowering relationships with local communities, this chapter focuses on examples that inspire us to consider radically different institutional-level relationships.

Expanding Ecoversities and Centering Community Needs

Universities are underleveraged institutions in society. Instead of continuing to center their own institutional reputation to survive and compete in a financialized world, imagine the impact that universities could have if they centered community needs instead of their own institutional needs. Just as chapter 5 imagines the societal impact of universities nurturing the generosity and loyalty of their students and alums toward vulnerable communities rather than back to the institution, here I propose that higher education systems could be restructured to center the needs of local and regional communities rather than centering their own institutional needs. One way to do this is to learn from, build on, and expand the experiences and principles of *ecoversities*. Ecoversities are people, organizations, and communities who are reclaiming knowledge systems and a cultural imaginary to restore and re-envision learning processes that are both personally meaningful and socially relevant to the challenges of our time (Ecoversities Alliance 2020).

The Ecoversities Alliance is a collective community of learning practitioners from around the world committed to reimagining higher education to cultivate human and ecological flourishing in response to the critical challenges of our time. The individuals, organizations, and programs that make up the Ecoversities Alliance are united by a shared exploration of what the university might look like if it were at the service of humanity's diverse ecologies, cultures, economies, spiritualities, and life on earth.

The expansion of ecoversities is part of an emerging knowledge movement that is building all over the world—a movement that has so far been largely ignored or unnoticed by most formal education systems. When I met one of the founders of the Ecoversities Alliance, Manish Jain, in India in 2018 (at Auroville's fiftieth anniversary event, discussed in chapter 3), I was immediately inspired and intrigued by the concept of ecoversities. Ecoversities are learners and communities who are co-creating and co-designing new approaches to higher education by reimagining diverse knowledges and relationships. By reconceptualizing and redesigning higher education to be inclusive of every human being and every human experience, ecoversities are initiatives seeking to transform the unsustainable and unjust economic, political, and social systems and mindsets that currently dominate global societies. This requires unlearning hierarchies of knowledge, unlearning hierarchies among people, and unlearning hierarchies among educational institutions.

Ecoversities are community-centered initiatives from every region of the world focused on a broad range of thematic areas, including healing, leadership, aquaculture, food, the economy, human rights, spirituality, and indigenous knowledge. Ecoversities include the Kufunda Village in Zimbabwe, a learning center where dozens of people live and work to co-create healthy vibrant community; the Swaraj Jail University in Udaipur Central Jail in India, a rehabilitative prison practice and learning program to reignite self-esteem, leadership, and life vision among incarcerated people; the Universidad del Medio Ambiente in Mexico, an institution offering multiple master's degree programs and a bachelor's degree all centered on regeneration and sustainability linking social and environmental systems, action research, and co-design; and Gaia U Latina, a living university without walls, crossing borders, like an un-institution, incubating a culture of ecosocial regeneration in Chile and Latin America. To appreciate the diversity among the hundreds of global initiatives that make up the Ecoversities Alliance, the website ecoversity.org is easy to navigate, with a global map of the ecoverse, descriptions, and links to each ecoversity, many publications, podcasts, and more. The six core values (table 3) embraced and nurtured by each ecoversity suggest a universality that higher education institutions around the world could also adopt to reconceptualize university-community relationships.

The first ecoversity value is *emergence*, which requires humility and letting go of illusions of control to allow for diversity. Second is *inquiry in solidarity*, which supports co-learning and compassion toward others' learning journeys. The third value is *experiential learning* as a way to connect our own experience with others', and the fourth is *emplacement*, which re-

Local Empowerment and Global Solidarity 195

Table 3 Values and Orientations of the Ecoversities Alliance

Values	Orientations
Emergence	An invitation to the unknown, allowing diverse ways of being, knowing, doing, relating to emerge
Inquiry in solidarity	An invitation to be authentic and critically engaged with co-learners, while invoking self-reflection, kindness, and compassion to support others in their own inquiries and discoveries
Experiential learning	Learning from our own senses, stories, spirits, hearts, hands, heads, and homes in order to find ways we are interconnected and entangled in each other's struggles and dreams
Emplacement	An invitation to reconnect with and learn from the land, the place, and the nonhuman; to engage in and promote deep localization
De-colonizing	An invitation to address, explore, and unlearn the dimensions of oppression, power, and privilege that are part of our own lives, relations, tools, structures, histories, and beliefs
Inter(trans)cultural dialogue	An invitation to learn in-between cultures, epistemologies, and cosmologies and to learn ways we might not recognize or have experienced before; to learn from/within/beyond diversity

Source: Ecoversities Alliance, https://ecoversities.org/.

fers to deep localization, learning from the land, the place, and the nonhuman. The fifth is *de-colonizing*, which requires unlearning oppression and privilege in our interactions, and the sixth and final principle is *inter(trans)cultural dialogue*, which is to learn from, within, and beyond diversity. Table 3 includes additional descriptions of the orientations associated with each of these values.

Among these six values, experiential learning and intercultural dialogue are frequently highlighted and embraced in many higher education institutions; these values are often provided as the rationale for supporting and encouraging community engagement. Decolonizing, emergence, and inquiry in solidarity are less frequently embraced because exerting power, exercising control, and acting individually remain central to how university-community relationships are currently structured in most places. Emplacement (engaging and promoting deep localization) is arguably the value that is most radically misaligned with many contemporary universities that have intentionally positioned themselves as distinct from their surrounding communities and the places where they are based. Most universities are structured to attract students from elsewhere and organized to prioritize their own survival and success, and all too often that is perceived as separate, and disconnected, from the well-being of the surrounding communities.

The legacy of colonialism is entrenched in the ways that universities theorize, research, and teach about "communities." Most universities perpetuate the colonial idea of "community-as-other." But universities redesigned and restructured with an explicit intent to serve the public good and advance climate justice could decolonialize the construct of community by embedding higher education institutions within the conception of community (Dutta 2018). While most universities support some community-engagement programs, and many promote initiatives to demonstrate that they are contributing positively to local communities, these efforts are often marginal compared to other institutional priorities. Some community engagement programs in higher education are strategically designed with a goal of promoting an appearance of engagement rather than achieving any specific community-centered goal.

Many communities have low expectations of local universities. Trusted, long-term relationships are needed for effective community partnerships, but relationship-building is often constrained. Many academics are trained within higher education institutions to center the university; it is difficult, therefore, to imagine a different kind of higher education system that would center community needs. The potential for co-designing and co-creating with community requires different models of university-community relationships and interactions. Creating new mechanisms for connection and dismantling conventional knowledge hierarchies are necessary to strengthen university-community relationships.

A society-wide commitment to emplacement and community-centered localization as a core mission of universities would require a geographic redistribution of higher education institutions because many communities are not near any universities. Just as many communities have access to local public libraries, this new vision for climate justice universities includes public higher education institutions distributed spatially so that all communities have easy access and proximity to university resources. Rather than being exclusive and exclusionary institutions, universities could be designed and supported to be all-inclusive hubs of civic engagement and action. Rather than being a drain on municipal finances (see chapter 5), universities could instead become a major asset, a public resource, for local communities. Rather than being spaces of private ownership and restricted access, climate justice universities could cultivate a sense of public space and open access, just as public libraries do. Reclaiming higher education for the public good could rely on restructuring networked university systems to be similar to, and perhaps partner with, the networks of public libraries that provide many resources, including books, magazines, internet access, and a diversity of programming, and serve as convening places, social hubs, and community centers in communities all around the world (Mehra and Davis

2015). While the well-funded and well-developed network of public libraries in the United States used to be a model for libraries in other countries, budget cuts, book bans, and political attacks have weakened this civic resource (Kahle 2023). The societal benefits of coordinated public investment in both public libraries and a linked network of distributed higher education institutions could be transformative by providing communities with a diversity of regenerative resources as well as learning and unlearning opportunities.

The current global distribution of universities is extremely uneven. Regional distribution of universities within countries is also uneven. To expand ecoversities and restructure higher education based on community-centered needs, a larger number of smaller localized publicly accessible universities is needed. As communities across the world are seeking information and support as they struggle to adapt to climate instability (Favretti 2023), a distributed, accessible public university system could create critically important community hubs providing learning opportunities and resources of all kinds.

This idea of reimagining a distributed system of public universities aligns with the proposal by the iconic social activist, environmentalist, author, and critic of globalization, Vandana Shiva, who made the case for creating grandmothers' universities everywhere. In the 2011 film *The Economics of Happiness*, Shiva says, "Local knowledge is knowledge that tells you about life. It is about living. I call it grandmothers' knowledge, and I think the biggest thing we need—the task for today is to create grandmothers' universities everywhere so local knowledge never disappears."

Like Shiva's proposal for distributed and localized grandmothers' universities, climate justice universities could establish networks of rural universities. Rural communities throughout the world are in need of reinvigorating investment and social infrastructure. A distributed network of public universities in all communities, including rural communities, would provide capacity-building support for communities to adapt to new climatic conditions and contribute to the co-design and co-production of new, emerging climate-related policies and practices contextualized to reflect the specific needs of each community and region. Climate justice universities could include a distributed network of rural universities serving as convening places to support learning by doing, peer learning, and experimentation of new approaches, including regenerative agriculture, renewable energy, regenerative forestry, and other regenerative land use practices.

In addition to reimagining a new geographic distribution of publicly accessible universities centering community needs, pressure continues to mount for existing universities to engage with local communities in new and different ways; careful attention must be paid to how this is done (Boyle, Ross, and Stephens 2011). As follow-up to his 2020 book *In the Shadow of the Ivory Tower: How Universities Are Plundering Our Cities* (Baldwin 2021), Davarian Baldwin has proposed that all existing universities conduct an iterative process of social footprint mapping (2022). Developed in Baldwin's Smart Cities Research Lab at Trinity College in Hartford, Connecticut, social footprint mapping involves making visible to the university community the interconnected tentacles of the institution that have an impact on the rest of the world. This includes mapping out the university's finances, real estate, technology transfer systems, development office, and research portfolio. Social footprint mapping is a way to create a culture of accountability within higher education and a way to begin the shift toward community-centered universities. The collective process of social footprint mapping supports collective action within the university and highlights how interconnected and embedded universities are with local communities. Social footprint mapping makes both the relationships and the abstract financial flows within a university more tangible and recognizable for all (Baldwin 2022).

Expectations for societal impact can be transformed if universities are restructured to center community needs. With climate chaos creating instability and exacerbating suffering in communities throughout the world, the opportunity ahead is for the infrastructure of higher education to be restructured to more directly address growing vulnerabilities.

Localism with Global Solidarity

The importance of localization, or the ecoversity value of emplacement, can be contextualized as direct resistance to the damaging trends of globalization. Globalization refers to the international integration of economies, societies, cultures, and technologies resulting in global networks of communication, production, trade, exchange, and finance (Sokol 2011). While the benefits of globalization are often characterized as expanded economic growth, enhanced job creation, improved access to information and resources, and facilitated cultural exchange and understanding, the numerous negative impacts include worsening economic inequities, financial instability, cultural homogenization, environmental degradation, and erasure of local traditions and identities (Cohen 2023; Farah and Junker 2021). The devastating impacts on the climate of deregulated market globalization were exposed in Naomi Klein's 2014 best-selling book *This Changes Everything: Capitalism vs. the Climate* (Klein 2014). After describing the disconnects between the extractive global economy and the transformation that is needed to confront the climate crisis, Klein calls for "a politics based on reconnection." Since then, climate suffering has gotten worse, and the climate injustices of globalization are increasingly recognized.

As extreme heat, fires, and smoke gripped the world in summer 2023 from Canada to Beijing, London, and Texas, the *New York Times* front-page headline on Sunday, June 18, read, "Failures of Globalization Shatter Long-Held Beliefs." The story reviewed the dire global economic realities of a war-torn, post-pandemic world experiencing climate chaos, highlighting the deep deficiencies in the assumptions of free market economics and pointing out that the market's invisible hand is not protecting the planet (Cohen 2023). As communities around the world struggle with both growing economic precarity and worsening climate vulnerabilities, the false promises of globalization are becoming clear to most. During this disruptive time, resistance to globalization is expanding.

One way to counter the devastating economic and environmental impacts of globalization is to support localism. Localism prioritizes citizens having greater control over the places in which they live and work and is often framed as a way to reinvigorate public engagement in democratic processes (Brownill and Bradley 2017; Wills 2016). The prioritization of local and regional interests promotes self-sufficiency by supporting policies, practices, and investments that encourage local production and consumption of goods and services, including food, energy, and materials (Brownill and Bradley 2017). Localism also focuses on local empowerment, including strengthening local governance, local culture, and local history. It also reconnects communities to the land, heritage, and ecology of the place. Given the massive and growing economic disparities between and among different localities (a pattern further exacerbated by globalization), localism, therefore, needs to be coupled with an inter-regional redistribution of wealth, power, and knowledge (Sokol 2003). Restructured networked higher education systems can be a key part of this process of spatial justice and spatial redistribution.

To leverage their transformative potential, higher education institutions can also contribute to the dynamic movement toward new municipalism (Roth, Russell, and Thompson 2023), as mentioned in chapter 5. New municipalism refers to a diversity of transformative initiatives and actions focused at the local level, including economic reorganizations, democratization of political decision-making, feminization of politics, and ecological transformation (Roth, Russell, and Thompson 2023). Universities embedded in their local communities could facilitate and engage with new municipalism ideas and initiatives by supporting an inclusive, participatory, and experimental approach to governance and by providing exnovation research (chapter 4) to resist and phase out neoliberal drivers that reinforce governance dictated by market logics for profit-driven politics (Sareen and Waagsaether 2022).

In addition to expanding a commitment to localism and new municipalism, climate justice universities can play a role in promoting our shared humanity by explicitly engaging in global solidarity (Smith 1998). English geographer David Smith challenged the social norm of caring more for those nearest and dearest to us and explored the possibilities of extending the scope toward a more universal care that relates to a more egalitarian theory of justice for all (Smith 1998). Prioritizing care and compassion, both near and far, in all aspects of university initiatives is fundamental to unleashing the potential of co-design and co-creation. Embracing a sense of global solidarity among all people and communities around the world is essential to resisting the "othering" and the dehumanization of some people that results from an increasingly financialized society full of lonely and disconnected people.

For universities to advance climate justice through relational knowledge and civic action, a new level of commitment to both localism and global solidarity needs to become a core mission of higher education institutions. To restructure and move away from the disconnected, individualistic, and competitive social norms that pervade contemporary higher education systems, an interconnected focus on both local empowerment and global solidarity anchors universities as places for interdependence, connectivity, and collective action. It is increasingly recognized that ideas, concepts, and opportunities for action are most tangible at interpersonal and local levels, and this holds true for institutions as well as individuals. How higher education institutions operate within the communities of which they are a part is central to defining their capacity for transformative civic engagement and impact. A place-based approach integrating teaching, research, and local or regional engagement is central to the visions of "the civic university" (Goddard et al. 2016), "the good university" (Connell 2019), the decolonized university (Bhambra, Gebrial, and Nisancioglu 2018), the engaged university (Beyond the Academy 2022), the plant-based university, and many other conceptions of universities designed for the public good.

Prioritizing local empowerment and global solidarity requires a paradigm shift within higher education because universities have gotten caught up in the assumptions of globalization. Many universities are themselves trying to globalize, establishing international campuses to expand into new "markets" in much the same way that multinational corporations do. This trend toward globalization in higher education has changed the way that some universities present themselves to recruit students. For example, Northeastern University, the US-based institution where I was on the faculty from 2016 to 2024, has multiple international campuses and now refers to itself as "a global university system" rather than a university based in Boston. Globalization of universities has been driven largely by the potential for institutional expansion and the tangible financial opportunities to universities in recruiting international students. The trend toward globalization has encouraged many universities to focus their communication and their partnerships to a nonlocal audience, which has in some cases reduced attention and relationship-building in the local area.

Reprioritizing local empowerment within higher education and restructuring universities to contribute to local and regional needs requires a different spatial distribution of university operations. Rather than continuing to support a great density of higher education institutions in certain urban centers while other regions have no universities, a physical redistribution of higher education institutions could be transformative. Investing in regionally distributed universities can provide multiple societal benefits to underinvested-in regions (Peer and Stoeglehner 2013). In addition to capacity building and economic benefits, regionally distributed higher education institutions strengthen community connections and reduce travel and commuting to centralized university hubs in big cities that can draw people away from their local region. Many countries are recognizing the value of decentralizing and distributing their university systems (Royal Irish Academy 2021). In Ireland, for example, the government has recently invested in the network of distributed rural technical institutes and created a new set of regional universities to better serve rural regions across the country.

When considering the role of universities in local empowerment, it is worth considering that the empowering role of universities is usually framed in terms of empowerment of individual students. Within this frame, higher education institutions influence society by facilitating student learning. The assumption is that when students complete their university degree, they will apply what they learned in their lives, in their jobs, and in their communities. It is increasingly recognized, however, that higher education institutions have multiple other mechanisms for empowerment, including empowering people and communities who are not enrolled as students. This broader framing of universities as communities of empowerment opens up possibilities to expand imaginations about the role of higher education in reducing inequities and vulnerabilities and addressing climate injustices.

With the global expansion of universities that has emerged in the past decade, that is, the opening up of new campuses in other parts of the world, the motivation for the new geographical locations has been largely based on financial returns. Universities are strategically selecting cities and regions where they expect the biggest returns on their investment, using a free market, extractivist logic. Rather than considering where in the world new campuses could have the biggest societal impact, these investments are made in new places to grow the institution. The vision and possibility of climate justice universities, however, would be to expand or create new offshoots in the regions of the world that have the greatest needs; this is a regenerative, justice-centered, caring-focused, solidarity approach—the opposite of the extractivist logic. This alternative regenerative approach is the distinguishing characteristic of all the initiatives within the Ecoversities Alliance.

Civic Engagement for Deliberative Democracy

Civic engagement is essential for deliberative democracy and a more equitable future. Civic engagement is also essential to address the rise in loneliness, disconnection, and growing mental health issues. Calls for a new civic economy (Dark

Matter Labs 2023), civics education (Allen 2023), sacred civics based on values (Engle, Agyeman, and Chung-Tiam-Fook 2022), and a civic revolution (Casale 2019) are growing as we all feel the consequences of outdated institutions and inadequate social and political infrastructures that are incapable of addressing the intersecting crises of our time. One approach to teaching, learning, and conducting research on civic engagement and a new civic economy is for higher education institutions themselves to expand their practice of civic engagement by establishing and sustaining new kinds of collaborative relationships outside of the university. If capacity building for deliberative democracy was a core mission of higher education, universities would necessarily create a different level of community-embeddedness in all of their educational and research activities. Participatory governance requires civic education, but in many places in the world, universities are not providing integrated civic education (Allen 2023). In the United States the decline in civic education has been attributed to the decline in democracy and a decline in pride of being American (Educating for American Democracy 2021). Civic renewal is needed in countries throughout the world to empower people to be involved and engaged and to facilitate communities taking back the power that corporate entities have acquired. Universities have a critically important role to play in civic renewal, especially if they are locally engaged and promoting global solidarity.

There are many ways to consider the role of universities in nurturing civic engagement. A classic model to distinguish among different levels and forms of citizen participation is the Arnstein ladder, developed in 1969 by Sherry Arnstein, a US-based public administrator (Arnstein 1969). Based on her experience working as a special assistant to the assistant secretary at the US Department of Health, Education, and Welfare, she developed insights on participatory decision-making, comparing different modes of citizen engagement. The bottom of the ladder represents the lowest levels of participation, including manipulation, which is considered pseudo-engagement because the participation is not actually contributing to the decision-making processes. The highest levels are "delegated power" and "citizen control," which represent participation processes when civil society input influences the outcome of decision-making processes. The middle rungs of the ladder include consultation and informing, which are both characterized by Arnstein as tokenism, which refers to when power holders restrict the input of citizens' ideas and "participation remains just a window-dressing ritual" (Arnstein 1969). Tokenism is evident when people are perceived as statistical abstractions and participation is measured simply by how many come to a public meeting, take brochures home, or answer a questionnaire. Arnstein points out that all that is achieved in this kind of engagement is that citizens have participated in participation, and powerholders have evidence that they have gone through the required motions of involving "those people."

In community engagement efforts at universities, the tendency for tokenized interactions is high. Arnstein's framework clarifies that for citizen engagement to be genuinely considered participation, it must allow for a redistribution of power. Without an authentic reallocation of power involving either redistribution of money or decision-making authority, participation processes are disempowering and reinforce the status quo because they allow the powerholders to claim all sides were considered without ensuring that all sides benefit. Civic engagement and university-community interactions can also be disempowering if an authentic redistribution of power is not assured.

Supporting Local Journalism

One key way that universities could contribute to a fundamental aspect of civic engagement and deliberative democracy is to empower local communities by strengthening local journalism. The loss of mainstream regional newspapers and the dearth of newspapers and other news outlets serving marginalized or under-represented communities around the world has resulted in a news desert crisis, with many communities having no access to local or regional news (Finneman, Heckman, and Walck 2022). An active and dynamic free press is an essential component for deliberative democracy (Sullivan 2020), so the lack of access to information and news regarding what is happening locally and regionally is not only disempowering and isolating, but it drastically narrows public discourse and public imagination based on whatever select news stories the national and global news outlets decide to include in their reporting. Research shows that people who live in news deserts without access to information about their communities are less likely to be actively involved in their community or participate in elections, and they are more likely to believe false information spread online (Gallup/Knight Foundation 2020). Given the financialization of the news media in countries throughout the world, national-level news reporting is heavily influenced by controversial and sensational stories that are likely to increase readership and newspaper sales (Silva 2014). Despite efforts by nonprofits, foundations, and some governments to increase financial support to bolster local media, news deserts are widespread, so there is large potential for universities to contribute and strengthen local journalism for the benefit of communities around the world.

Multiple creative initiatives provide examples of a new model of academic engagement and impact in local media. *The Scope* is an experiential digital magazine that covers local news, produces watchdog journalism, and seeks diverse commentary and in-depth analysis of equity and justice issues focused on Roxbury, one of Boston's historically Black neighborhoods that is often left out of mainstream Boston-area news. This media outlet is operated by Northeastern University's School of Journalism and staffed by undergraduate and graduate students who are supervised by a full-time professional editor and a faculty advisor (Finneman, Heckman, and Walck 2022; *The Scope* 2023). Elsewhere, *the Eudora Times*, an online community newspaper for the small town of Eudora, Kansas, began as a one-semester class project within a Kansas University social media class in 2019; the town of just over six thousand residents had lost its local newspaper in 2009 and had been a news desert before the students began reporting on the school board, the city commission, and public health meetings as well as writing feature stories on community residents, local businesses, high school sports, and local churches (Finneman, Heckman, and Walck 2022). These examples show the transformative power of university engagement in local journalism; the transformative vision of climate justice universities can leverage this transformative potential by providing a distributed journalistic presence in all communities.

Given the links among knowledge, wealth, and power discussed in chapter 1, when universities contribute to local journalism, they are empowering communities with local knowledge, equipping them to engage with local resources and local wealth regeneration. Local journalism can strengthen community connections and networks to promote local coalitions to advocate for local initiatives and regenerative economies in the region.

Community Partnership for Transformation: The Case of DRIFT

An illustrative example of a university initiative that is co-creating knowledge to empower communities toward regenerative transformation is the Dutch Research Institute for Transitions (DRIFT) at Erasmus University Rotterdam (also mentioned in chapter 4). Cooperation and partnership are recognized as critical aspects of designing higher education for regional sustainable transitions (Hoinle, Roose, and Shekhar 2021), and DRIFT demonstrates the potential for much larger societal impact with an alternative, innovative model of university-community interaction. Established in 2004 as an interdisciplinary faculty group, DRIFT focuses on practical applied sustainability transition projects and co-production of knowledge with local governments, nongovernmental organizations, policymakers, communities, and business actors that are looking for ways to move beyond business as usual. In its early years it gained a strong reputation within the Netherlands and beyond for innovative, engaged sustainability transitions work, but because it did not fit the traditional model of an academic unit, in 2011 it was ousted from the university as an academic entity and became an independent company (still under the holding of the university). DRIFT then had to develop its own novel financial model to sustain itself, and over the next decade it became extremely successful in co-designing for sustainable transitions, embracing a plurality of definitions of sustainability, and working with the assumption that current social practices and structures are perpetuating unsustainability (Wittmayer and Loorbach in progress).

Within the new independent model, DRIFT was no longer constrained by the conventions of the university. It rapidly expanded the number and types of community partnerships and organizational projects while maintaining research output and external grant funding that exceeded expectations of academic units within the university. Through facilitated knowledge co-production with a diversity of nonacademic actors engaging with sustainability transition applications, DRIFT emphasizes the continuous process of reflecting on meaning and exploration through applied experimentation. Issues of power and governance regularly permeate their projects, so DRIFT researchers were able to expand the academic literature on power dynamics (Avelino 2017, 2021; Avelino and Rotmans 2009) and critically question assumptions of apolitical or neutral positions by other scientists, experts, and stakeholders (Wittmayer and Loorbach in progress).

The DRIFT approach combines interdisciplinary academic research with co-creative participatory processes through which analyses are validated, enriched, and translated into everyday contexts. Rather than attempting to predict or control the future, they collectively embrace emergence, experimentation, and learning by doing (Wittmayer and Loorbach in progress). Recognizing uncertainty and acknowledging that transformative change is inherently ambiguous and contested, they anticipate all kinds of psychological, institutional, economic, and societal emotions and forms of resistance and conflict. In contrast to dominant academic structures that are output oriented, disciplinary, research centric, and incremental, DRIFT sees engagement with pioneering practice-based contexts as experiential social science in which practical contexts are used to test hypotheses and experimentally develop and refine new methods, models, and interventions (Wittmayer and Loorbach in progress). Generic and robust outcomes are then translated to academic funded research projects and validated through academic publications. This reversed validation process is distinct from how social science research is institutionalized at most universities (Wittmayer and Loorbach in progress).

By all academic measures, DRIFT became even more successful when it was released from the constraints of academic institutions. With regard to community engagement and societal impact, its expanding reputation and sphere of influence validates the effectiveness of this community-centered alternative approach. After more than a decade of this expansive impact, the university became increasingly interested in the DRIFT approach and invested in a university-wide "institutional experiment" called the Design Impact Transition (DIT) platform to try to integrate transdisciplinary, collaborative, and action-oriented academic work that explicitly aims to support sustainable transitions throughout the entire university (Loorbach and Wittmayer 2024). This inspiring example of an existing alternative model provides valuable insights for restructuring climate justice universities.

Think Globally, Act Locally

The phrase "Think Globally, Act Locally" emerged in the environmental movements of the early 1970s to encourage individuals and communities to engage with global environmental challenges through local action. As the world becomes increasingly volatile and disruptive, leveraging and expanding on this concept is an essential component of reimagining how to leverage the transformative potential of higher education institutions. This sentiment is reiterated in many contexts by people with different agendas and concerns. For example, the call to think big while simultaneously thinking small is a key message made by J. K. Gibson-Graham and colleagues in their 2013 book *Take Back the Economy: An Ethical Guide for Transforming Our Communities* (Gibson-Graham, Cameron, and Healy 2013). These authors and multiple previous and subsequent scholars and practitioners reinforce that idea that societal transformation requires empowerment at the local level, including small-scale practices and interactions with the people and communities close to us, while also connecting in solidarity with larger global struggles. To reclaim higher education for transformative change for justice, stability, and hope, a restructuring of institutional priorities toward local empowerment and global solidarity is necessary.

This call for universities to orient their institutional priorities toward thinking globally and acting locally is aligned with efforts by many universities to reduce academic air travel emissions by encouraging students, faculty, and staff to travel less. Recognizing that air travel is one of the most carbon-emitting activities that individuals can undertake and acknowledging that aviation contributes 4%–5% of global greenhouse gas emissions, reducing air travel has become a critical issue for universities committed to ambitious emission reductions (Tseng, Lee, and Higham 2022). Alternatives to flying vary widely depending on location and context (Thaller, Schreuer, and Posch 2021); while academics in much of Europe have access to well-developed extensive train networks, many other parts of the world and islanded nations have more limited options.

Changing the incentive structures toward local engagement and reducing the perceived competitive need to fly around the world to achieve academic success would contribute to reducing universities' contributions to greenhouse gas emissions (Thaller, Schreuer, and Posch 2021). It is also important to point out here that a focus on individual actions and individual decisions on whether or not to fly shifts responsibility away from the powerful aviation industry and the fossil fuel industry that is investing heavily to perpetuate and reinforce expectations for low-cost air travel. Flight shaming and other strategies that place blame on individuals detract from the larger structural changes that are needed. Within the current systems, many students and academic staff do not feel like they have much choice in how and whether they travel; if universities want to encourage alternative modes of transportation (including boats, trains, cycling, etc.) then incentives and support to empower people to make alternative choices need to be integrated.

With regard to student mobility, many universities still rely heavily on recruiting international students, and most universities continue to encourage and facilitate study abroad experiences and educational travel. For centuries, travel and exploration have been closely linked to the privileges of higher education, so for students who can afford to travel, it may seem challenging to reduce this or to use alternatives to flying. If universities are restructured to be more locally embedded and community connected, fulfilling, transformative opportunities within the local community would be expanded both for current students and for alumni. Students may be more likely to stay in the local region after graduation if they feel grounded in and connected with extended social networks.

There are many potential benefits and emergent opportunities associated with universities prioritizing and incentivizing thinking globally while acting locally. With the weakening social fabric that weaves people together and keeps communities connected, locally focused universities could provide a diversity of valuable social infrastructures accessible to all. The possibilities are unlimited and range from universities hosting citizen assemblies on controversial political issues, to partnering with local food providers to expand local distribution, to co-creating community-centered workshops for how to start up a cooperatively owned business.

As climate chaos exacerbates vulnerabilities, the potential

and need for universities to collaborate, empower, and support local and regional communities is expanding rapidly. To uphold deliberative democracies and the social infrastructures required for deliberative democracies to thrive, a paradigm shift in what society expects from universities can be imagined. Instead of continuing to cater to corporate interests and preparing students to work for financialized global companies for profit, institutions of higher education can expand their commitment to civic engagement, deliberative democracy, and localism, coupled with a strong sense of solidarity with people around the world. As the growing precarity of humanity becomes clearer and more obvious, universities have an opportunity to expand their global impact through international solidarity while simultaneously engaging with the renewed centrality of local, democratic public processes.

Most institutions of higher education are anchor institutions, meaning they have a long-term, significant presence and influence on the land and in the communities within the geographic area where they are located. Unfortunately, the financialization of higher education has minimized genuine institutional commitments and collaborations to local land and local communities and replaced them with an extractive approach. In the United State in particular, universities have leveraged their tax-free status as nonprofit institutions to buy land and expand their campuses. As mentioned in chapter 5, this trend of expanding universities' ownership in many cities has had devastating impacts on local communities; gentrification has caused displacement and forced many families, households, and even whole communities to move (Baldwin 2021). Restructuring regenerative, reciprocal, empowering relationships between universities and local communities is fundamental if universities are to unleash their potential to advance climate justice.

A reimagined networked system of universities could be a key component of building climate resilience in vulnerable places around the world. The cross-university collaboration, the RISE Network mentioned at the beginning of this chapter, that emerged after Hurricane Maria devastated Puerto Rico demonstrates the potential for coordination and solidarity among higher education institutions. Unfortunately, due to the colonial, capitalistic context of the recovery and reconstruction in Puerto Rico and a lack of public investment from the United States government (De Onis 2021; Kuhl, Stephens, et al. 2024), the suffering, deaths, worsening economic precarity, and extensive migration away from Puerto Rico have further marginalized communities who were already vulnerable. The violent suffering of climate coloniality will continue in Puerto Rico and many other places until and unless there is transformative economic and political change toward climate justice (Kuhl, Perez-Lugo, et al. 2024). Consistent with the vision of climate justice universities, the University of Puerto Rico, including its eleven distributed campuses, could be reinvested in to become critical infrastructure to serve the needs of Puerto Rican communities. Building on the community-centered leadership of the RISE Network, the shared principles of the Ecoversities Alliance, and the co-production approach of DRIFT, climate justice universities require restructuring and redistributing higher education institutions to prioritize local empowerment and global solidarities to advance transformative climate justice.

CONCLUSION

Toward Climate Justice Universities

Recognizing the power and influence of higher education in shaping our collective futures, the injustices of the climate crisis justify a creative reimagining of the value and purpose of universities. The idea of climate justice universities represents a paradigm shift away from academic frameworks that promote narrow, technocratic, individualistic "solutions" toward more expansive and regenerative academic engagements to facilitate structural and systemic transformative change.

I first proposed the idea of climate justice universities to an international audience at a public lecture that I gave in November 2023 at the Harvard Radcliffe Institute for Advanced Studies, where I was a Climate Justice Fellow for the 2023–2024 academic year. At this time, the Harvard campus, like many universities in the United States, was embroiled in public controversy related to its perceived bias toward those supporting the Israeli government's response to the brutal Hamas attacks of October 7, 2023, and its repression of free expression among those advocating for peace and liberation for the Palestinian people. As wealthy Zionist donors pressured the university with threats to withdraw financial support, the university took additional disturbing steps to further suppress expressions of solidarity with Palestinians. As I prepared my public lecture within this tense campus environment, I recognized the need for courage. To present my creative critique of higher education institutions at Harvard, one of the wealthiest and most powerful and influential universities in the world, was both counter-hegemonic and counter-cultural. Discussing how wealth and power are constricting academic inquiry and distorting knowledge creation and knowledge dissemination at universities seemed both timely and risky. In these challenging circumstances, I felt honored to have this unique opportunity to share my ideas by speaking truth to power.

The responses I received after this public lecture fit into two broad categories: appreciative enthusiasm and resistant skepticism. The presentation was livestreamed (and recorded and subsequently posted on the Radcliffe Institute website*), so I received feedback from people around the world. In addition to appreciation from environmental justice–oriented peers at multiple different universities, people engaged in climate justice activism, feminist scholarship, racial justice, and human rights were among those who offered the most enthusiastic feedback. The skeptical responses were predominantly from older academics, mostly men—one man asserted with great authority that universities are businesses, so we should never expect them to have a social justice mission.

While I had assumed my ideas about climate justice universities might have generated interest and engagement at Harvard, which had funded my climate justice fellowship to develop these ideas, only three Harvard faculty offered responses to my talk. An economist who attended was skeptical and suggested that I should provide more concrete data and evidence to demonstrate the connections that I was asserting. A social scientist who studies science and technology congratulated me by saying that I had "hit the nail on the head," and a public policy professor was impressed with the way that I integrated

---

*Presentation available at https://www.radcliffe.harvard.edu/event/2023-jennie-c-stephens-fellow-presentation-virtual.

so many provocative ideas. Despite Harvard receiving a $200 million donation in 2022 to expand its work on climate and sustainability, an administrative leader at Harvard confirmed to me a few months later in a subsequent meeting that the institution is too conservative to focus on societal transformation.

Among the most inspiring responses I received after my public lecture was from a professor of education from Jawaharlal Nehru University in New Delhi, India, who is actively engaged in knowledge co-production of climate science with locals in the Ladakh region in Himalaya. He watched the livestream and wrote me an email directly afterward in which he thanked me for my thought-provoking talk and shared with me his interest in expanding university space as a new ecology for pedagogy and reflexive action. We have since communicated several times, and we are planning to collaborate and share insights from our experiences.

The Power of Legacy

The idea of reclaiming and restructuring universities using climate justice principles relies on the understanding of, and building on, the legacy of higher education institutions. Each university and every higher education system is embedded within the context of its own institutional history. Recognizing the power of legacy, during my time as a visiting scholar at Harvard I was intrigued to learn about the Harvard Legacy of Slavery Initiative, so I signed up for the walking tour that shares some of the history. The tour typically includes ten different notable locations around the campus, but because of student protests in support of Palestine in Harvard Yard that day, the tour was shortened to include only six places. The first stop, a large four-column building in Radcliffe Yard, was the Elizabeth Cary Agassiz House, which is currently the Harvard College admissions visitors center, where prospective students and their families begin tours of the university. This building is named in honor of Elizabeth Cary Agassiz, who was a staunch advocate for women's education and the cofounder and first president of Radcliffe College, a women's college established in 1894 in association with Harvard. She was also the wife and scholarly partner of Louis Agassiz, the Harvard professor mentioned in chapter 2, who was an influential advocate of white supremacy and racial segregation. To fund his research and advocacy on polygenesis, which is the discredited and harmful idea that different races are different species so there was a hierarchy among different races, Professor Agassiz raised money from elite networks to fund his research and to establish a museum at Harvard to popularize his science and make it more accessible. The third stop on the tour was the Peabody Museum, within which there are still human remains of indigenous and enslaved people; one of the museum buildings also has the name Agassiz prominently engraved on the front to honor Louis Agassiz.

While the Legacy of Slavery Initiative acknowledges and reveals the power and influence that Louis Agassiz had toward legitimizing white supremacy and racial segregation in the United States and around the world, it is also worth remembering that Agassiz's views were widely accepted at the time. His position as a scientist and professor at Harvard brought credibility and respect to these ideas; he used the power of the university to promote certain kinds of ideas that perpetuated economic exploitation and dehumanization of Black and other non-white people.

One key part of the legacy that seemed to be missing from the tour was exploration into how current researchers and professors at Harvard are using the power and prestige of the university to promote certain kinds of ideas that perpetuate continued economic exploitation and dehumanization of some people. The assumption seems to be that Harvard's racist, patriarchal, colonial practices and policies are a thing of the past. My professional experiences at Harvard suggest that this is not the case. Academic frameworks at Harvard and many other universities continue to promote disconnection and dehumanization that erode a sense of common good and reinforce complacency to human suffering. Silencing of certain people and suppression of certain ideas are accomplished through various forms of subtle and blatant intimidation and through financial structures.

Within Harvard's community of scholars focused on climate and sustainability, not only is there minimal interest in engaging with transformative climate justice, but the dialogic culture among the environmental science community is characterized by an ultracompetitive, arrogant, toxic masculinity that excludes certain voices, dissuades many perspectives, and constrains who engages in community-wide discussions. After attending a few uncomfortable large events where a series of disrespectful, attacking comments were made publicly to the person presenting, I decided to join others that I know (both junior scholars and full professors) who have strategically decided not to participate in these community events. I recognize the paradoxical realities of my decision; when the climate justice fellows at Harvard opt out of attending the community-wide events on climate, the mainstream (and malestream) "climate isolation" approach to climate will not be challenged. This is why the vision of climate justice universities requires a paradigm shift and a transformative change in existing power structures.

The need for a paradigm shift in higher education is also evident from how wealth and power in US universities are undermining the legitimacy of university leadership. Among Harvard's senior leadership, a series of high-level decisions responding to campus protests demonstrates a different, but aligned, intimidation and suppression of certain ideas. Sustained efforts to vilify those on campus who advocate for human rights and speak out against the oppression of Palestinian people demonstrate how white economic privilege continues to buy institutional protection. The right to study and critique state violence is fundamental to academic freedom: if that right is not protected within universities, the independence of higher education institutions is gone; they may as well become part of national militaries. The urgent need for a paradigm shift in higher education is clear when university leaders succumb to financial and political pressures to use tactics of intimidation and harassment to discourage and dissuade expression of certain views.

Beyond the examples here from Harvard, the power of legacy is situated in all contemporary universities. In reimagining the paradigm shift toward climate justice universities, the goal is not to erase or dismiss legacy but to creatively build on the legacy in regenerative and reparative ways. While there is much to critique within the Harvard Legacy of Slavery Initiative, the conceptual commitment to acknowledging harm and then opening up community dialogue about reparations are important intentional steps that take courage and care. A core principle of climate justice is acknowledging harm. For higher education, this requires promoting critical self-reflection not just of past harms but also on how current and future institutional decisions contribute to current and future injustices. Acknowledging harm also requires a courageous institutional commitment to transparency, including financial transparency, transparency about internal governance as well as transparency in all interactions with external communities.

Transformation across and beyond Individual Universities

While much of the exploration in this book has focused on imagining change within individual universities, the transformative potential of higher education requires networked consideration and co-creation across higher education institutions as well as co-design with individuals, organizations, and policymakers outside of the sector. The paradigm shift toward climate justice universities requires leaders at every level within and outside higher education to pay attention to new ways of leveraging the power and influence of universities for a more just, equitable, and healthy future. A networked distributed approach will expand access and impact for transformative change. Just as networks of public libraries coordinate to provide knowledge resources for distributed communities in almost every country of the world, expanded distributed networks of universities have potential to facilitate knowledge co-production to reduce climate vulnerabilities and promote community well-being in communities throughout the world.

In 2020, it was estimated that there are over 28,000 institutions of higher education in the world, with about 250 million students; one estimation projects growth in the number of students to about 600 million by 2040 (Lueddeke 2020). Among the 28,000 institutions, most are under-resourced and financially struggling. More equitable distribution of resources and funding for higher education institutions is a policy priority for climate justice transformation. To reduce human suffering, every community and region needs some access to universities so that knowledge can be co-created and contextualized for the challenges facing diverse local communities around the world.

Given the scale and scope of the global higher education sector, more intentional global solidarity among higher education institutions and individuals is possible. Collective action and coordination among these organizations increases the power and influence of the sector, so new networks, linkages, and mutual support among and between individual universities is needed. The International Association of Universities (IAU) is an existing global organization that represents and advocates for the higher education sector in global governance. Created by UNESCO in 1950, IAU is a membership-based organization that serves the global higher education community (over 550 member institutions) by providing advice, trends analysis, and publications to its members and facilitating connections and peer-to-peer learning among institutions, convening events and advocating for higher education in various contexts.

One challenge with the IAU model is that because the organization is membership based, universities with fewer resources are unable to join so they are less well represented in the global landscape. At an international conference on higher education reform in Glasgow, Scotland, in June 2023, I met the secretary general of the IAU, Hilligje van't Land from France, who during her opening keynote presentation described the important role IAU plays in advocating for more intentional consideration of the value of higher education in international policy and global governance and how IAU tries to decenter European universities in their work and elevate and expand the global network. She mentioned how universities in the United States are among the least engaged in the global network, in large part because they are self-absorbed in the large and complex national higher education landscape.

Multiple other international networks exist to connect individuals who are working toward transformation within and beyond universities around the world. These include the Ecoversities Alliance, a global networked collective of learners and communities reclaiming diverse knowledges, relationships, and imaginations to explore what the university could be and designing new approaches to higher education that serve diverse ecologies, cultures, economies, and spiritualities (introduced in chapter 3 and discussed more in chapter 6); and Faculty for a Future, a community of academics who feel a duty of care over earth's intersecting social and environmental crises. Their work champions academic practices that embrace and embody the crucial need for societal transformation. This network has developed a "people-powered university" initiative to support participatory assemblies where facilitated dialogue within university communities can occur. Scientist Rebellion is a network of scientists and academics who expose the reality and severity of the climate and ecological emergency by engaging in nonviolent civil disobedience. Scientists for Global Responsibility is an independent UK-based membership organization of natural scientists, social scientists, engineers, IT professionals, and architects who provide a support network for ethically concerned professionals in these fields and beyond. And Planetary Limits Academic Network (PLAN) is a decentralized network of scholars across disciplines committed to collectively addressing critical systemic challenges facing humanity by fostering a radically interdisciplinary approach to creating responses grounded in planetary and ecological limits. Scholars for a New Deal for Higher Education (SFNDHE) is a US-based advocacy organization of educators calling for rebuilding higher education for the public good in a way that works for everyone by increasing federal investment and reducing precarious workers and student debt.

Educational policy at national and regional levels has significant influence over how individual universities are managed, regulated, and financed. Policymakers' perceptions, priorities, and assumptions regarding what can be expected from higher education institutions and how universities can be leveraged for the common good determine funding levels and regulatory constraints and incentives within the higher education sector. Reclaiming and restructuring higher education systems, therefore, requires political engagement from individuals beyond those that currently work or study at universities. How national governments prioritize investment in higher education is critical to defining their capacity to engage in new ways. Despite the critical importance of public policy and political landscapes, a contextualized discussion about higher education policy in individual countries is beyond the scope of this book.

To further support and expand global solidarity in the movement to reclaim and restructure higher education systems throughout the world, additional transnational advocacy and international networks supporting the principles of climate justice in higher education are needed. Regionally, bottom-up collaboration and coordination, like the convening of over one hundred academics and activists in Galway in Ireland mentioned in chapter 1, can build a sense of collective action and shared community with a focus beyond individual universities. Transformative change requires synergistic pressure at multiple levels, so local, regional, and global networks exploring opportunities for change are all needed, valuable, and mutually supportive.

Purpose, Accountability, and Values

Much has been written about how universities are struggling to define their purpose; in his 2020 book *Learning-Centered Leadership in Higher Education* Ralf St. Clair highlights "the muddled missions" of universities. From rising in the rankings, to creating impact, to fostering innovation, to providing an elite finishing school for youth, to prioritizing community engagement and experiential learning, there are a multitude of competing claims being made by universities about their mission and what they are trying to do. Given these tensions coupled with expanding skepticism of the value of an expensive financialized university education, now is an appropriate time for higher education institutions to clarify their purpose and consider a renewed commitment to climate justice and shaping a hopeful future for all.

Within the private sector, many companies around the world have redefined themselves as "purpose-driven organizations" committed to aligning with societal interests and reorienting their mission with a sustainable future; a similar shift has not been seen in higher education (Hurth and Stewart 2022). While many universities claim to be contributing to sustainable efforts, most sustainability initiatives and commitments in higher education are incremental and nontransformative and fail to integrate into high-level strategic decision-making to fulfill the core purpose of the institution. The inadequacy of climate and sustainability efforts in higher education is increasingly acknowledged, and pressure is mounting for universities to do more (Kinol et al. 2023). Universities can no longer afford to be agnostic on issues of climate justice, and a genuine deep institutional commitment to acting to advance climate justice has potential to provide a core purpose that guides universities' strategic efforts (Kinol et al. 2023).

For climate justice to become an expansive and inclusive values-based purpose guiding transformation within higher education, a culture of care must be nurtured. Although the patriarchal and colonial legacies of universities have dismissed the importance of care and compassion, devalued the importance of care work, and erased the centrality of relational knowledge (i.e., how we are all connected to each other and to the nonhuman world), a shift toward a culture of care is a central part of transformative climate justice. If higher education institutions were restructured to prioritize both human health and ecological health, transformative ideas like the solidarity economy (Matthaei and Slaats 2023) and economies that "dare to care" (Lorek, Power, and Parker 2023) would be prominently featured. The case is increasingly being made by a diversity of economists, feminists, and social justice advocates that centering care is essential to promoting social justice and preventing ecological breakdown (Lorek, Power, and Parker 2023; Oksala 2023). Care-centered economic change could be the holistic narrative and approach needed for powerful systemic change; there is a growing global movement to promote care as a catalyst for radical transformation (Lorek, Power, and Parker 2023). A commitment by universities to orient their programs and initiatives toward caring societies is another purposeful framework for guiding priorities within higher education.

In her 2023 book, *Catalyzing Transformation: Making System Change Happen*, management scholar Sandra Waddock identifies five core dimensions of systemic change: *purpose* is the core reason for the existence of a given entity or system; *paradigms* (also referred to as mindsets, mental models, or perspectives) are the beliefs or framings that provide a narrative that helps actors situate themselves within the system; *performance metrics* are the metrics, assessment, and evaluation criteria that get recognized and rewarded within the system; *power relations* are the formal and informal organizational arrangements and structures that determine who has access to resources and who doesn't; and *practices* (including policies, procedures, and processes) include all the different ways that work gets done so that the purpose of the system is fulfilled. These five dimensions are interconnected: change in one dimension can trigger change in another. Applying these dimensions to catalyzing transformation in universities allows us to recognize that there is not one leverage point, but multiple possibilities for synergistic change. Similarly, there are multiple and varied obstacles to change. While much of the discussion in this book has focused on a paradigm shift and envisioning a clearer purpose, shifting performance metrics, power relations, and dominant practices have also been explored.

The vision of climate justice universities requires reclaiming a culture of accountability and transparency within higher education. To sustain legitimacy and trust, there should be no hidden agendas or anonymous donors. Full disclosure about finances as well as transparency in all institutional strategic decisions are essential for intellectual honesty within reflexive organizations that are committed to institutional learning. A lack of attention to creating cultures of accountability in universities has been identified with concern for how this impacts equity and inclusion efforts (Shaibah 2023). Like many corporations, universities are increasingly engaged in greenwashing (Cownie 2021), which refers to misleading or deceptive claims about sustainability initiatives (Nemes et al. 2022). Growing awareness of the prevalence of greenwashing in the private sector has raised skepticism throughout society; cynicism among students and staff regarding the claims their universities are making about their sustainability initiatives is common.

With colleagues, climate action scholar Paul Lachapelle explores academic capture in higher education climate action, the co-option of climate action language in ways that are not actually transformative (Lachapelle et al. 2024). Given that transparency and accountability are essential for learning organizations and learning-centered leadership (St. Clair 2020), the lack of attention to creating cultures of accountability in contemporary universities is problematic and contrary to learning principles. Recognizing that how we organize our institutions is interwoven with how we perceive ourselves, a reimagination of universities requires critical self-reflection.

Imagination for Justice

Envisioning a different future with the power of imagination is an essential part of social change and the struggle for justice. At the 2023 University of Southern California graduation ceremony, sociologist Ruha Benjamin reflected on the critically important role of imagination in the struggle for justice:

> The more steeped I am in the facts of inequity and injustice, the more I come to appreciate the importance of imagination as a field of struggle. Not an ephemeral afterthought that we have the luxury to dismiss or romanticize. Social imagination, moral imagination, decolonial, anti-racist, collective imagination. Remember our laws and policies, our systems and structures are not timeless. They all began at some point with how people imagined how things could be, how things should be.

Benjamin, who expands on the centrality of imagination in her 2024 book, *Imagination: A Manifesto*, highlights the empowering role of our imaginations. While contemporary universities tend to promote and prioritize a technocratic and linear way of thinking about the future, imagining alternative futures is central to the work of building a better world for all.

In the same speech at the University of Southern California, Benjamin also defined ignorance and knowledge in terms of resisting and advancing social change: "Ignorance is not simply a lack of knowledge—but a distorted lens that is actively produced by those who seek to maintain business as usual. If ignorance is an industry then what is our job as educators?" Here she is referring to the power of knowledge and the strategic power of those who manipulate and restrict certain kinds of knowledge to resist change. Agnotology, the study of ignorance—a field that explores why we don't know what we don't know—reminds us that ignorance can emerge from cultural and political struggles, and it often results from strategic efforts to prevent people from knowing certain things (Proctor and Schiebinger 2008). With the commercialization of research and the close relationships between industries and universities, tension and controversy regarding whether and how universities are contributing to or complicit in ignorance-constructive practices have emerged (Pinto 2015). Acknowledging the multiple ways that many higher education institutions constrain learning and restrict knowledge production and dissemination, Benjamin's commencement speech legitimizes the value and importance of creating intentional academic space for imagining.

Instead of reinforcing the narrowly conceptualized linear logic that dominates university innovation activities, one that relies on defining a discrete problem and then developing solutions, Benjamin is endorsing a very different model for academic inquiry. She is suggesting that universities, educators, and researchers nurture the imagination to allow for more radical and transformative thinking. The shift from a mindset focused on solving problems to an approach that prioritizes creating possibilities is a fundamental change. Rather than constricting our thinking by defining narrow problems that need to be "solved," universities can be reconceptualized and reinvested in to ensure they are creative spaces that encourage people and communities to co-create alternative possibilities.

Black feminists have a long tradition of imagining alternative worlds, a practice that is fundamental to the struggle for freedom and justice. As American science fiction novelist and award-winning author Octavia Butler wrote in her unpublished manuscript "Parable of the Trickster," "There is nothing new under the sun but there are new suns" (Canavan 2014). Living in a world defined by structural oppression and systemic carcerality requires imagining a different world not just for individual futures but also for our collective futures. As American civil rights leader Fannie Lou Hamer famously said in a speech at the founding of the National Women's Political Caucus in 1971, "Nobody's free until everybody's free."

Black feminist imagination and the practice of Black feminist worldmaking are relevant to all who are working toward a just, climate-stable future through education. The Black feminist scholar, activist, and professor bell hooks (whose inspiring 1994 book *Teaching to Transgress: Education as a Practice of Freedom* is mentioned in chapter 3) was committed to making the ideas of Black feminism accessible to all. In her 2015 book *Feminism Is for Everyone*, hooks encourages readers to "imagine living in a world where there is no domination . . . living in a world where we can be who we are, a world of peace and possibility." In this short accessible introduction to feminist theory, hooks demonstrates the power of Black feminist imagination and how it is an essential part of creating hope and possibility and resisting the realities of the world as it is now.

In too many contemporary universities, space for imagining alternative futures is constrained. The tech-focused, engineering-oriented approach of defining a problem and then narrowly developing a solution to that problem continues to dominate how higher education institutions are conceptualizing their impact on society. Integrating the wisdom, theory, and practice from Black feminists and other communities and traditions that have been systematically marginalized in higher education institutions into the policies, priorities, and practices of universities is a necessary part of the paradigm shift toward climate justice universities.

At the inspiring 2023 convening of academic staff and activists in Galway, Ireland, focused on how higher education should respond to the planetary crises, opening up space for imagining transformation emerged as a key intention for the subsequent meetings of the network. Co-organizer and co-convener John Barry suggests that a goal for all of the meetings in each of Ireland's four provinces is to also visit the most important province we need to discuss transformative change—that is the "fifth province; of ancient Irish mythology, the place of creativity and imagination" (Barry 2023, personal communication). Imagination is indeed crucial for taking a transformative leap—and while making the leap we should remember the words of South African anti-apartheid activist and politician Nelson Mandela: "It always seems impossible until it's done."

This perception of impossibility is fueled by powerful interests who defend the status quo because they are unable to imagine alternatives. Just as many powerful people and institutions defend capitalism by reinforcing the narrative that there is no possible alternative economic structure, many powerful people and institutions in universities defend the financialized model of higher education by reinforcing the narrative that there are no possible alternative structures. But as the inadequacies, inequities, and injustices of so many higher education systems throughout the world are increasingly revealed, taking time to imagine alternative models is an act of hope and a move toward justice.

Universities as Critical Infrastructure

The vision of climate justice universities in this book is based on the premise that higher education institutions are under-leveraged resources for society. I make the case that higher education, if structured and financed differently, could have a much more expansive, positive societal impact advancing the public good than it currently does. One way to conceptualize and justify a transformation in higher education is to consider universities as critical social infrastructure that needs to adapt to the dynamic changing times. Critical infrastructure refers to the systems, assets, and networks that are essential for the functioning of society (Ali 2021). Energy, transportation, health care, and communication are all considered critical infrastructure. For civic engagement in a time of rapid change, accessible, distributed public education for all is also essential infrastructure. The possibility of linking restructured higher education systems with public libraries and public schools, infrastructure that is already more widely accessible and equitably distributed than universities, provides a tangible vision to build toward.

That dictators often try to shut down or control universities highlights their power and social value (Douglass 2021). Within the first four months of Israel's war in Gaza, all twelve of the universities in Gaza were destroyed. Attacking universities, schools, and cultural sites is recognized as a part of genocide because eliminating culture and knowledge of a people is a way to eliminate those people (Desai 2024). Understanding how universities are manipulated (or destroyed) in political conflicts and colonial contexts offers insights into the societal impact of higher education institutions. While restraint and cautious consideration of political sensitivities are necessary for the survival of some universities and some individuals within universities, the purpose of a commitment to academic freedom is to protect scholars so they can advocate for change and challenge dominant beliefs and assumptions.

At the Galway event in Ireland, one powerful intervention was made by Hannah Daly, a professor and scholar on energy system change. She spoke about academic freedom as both a privilege and a duty while reminding everyone how Irish law defines academic freedom: "A member of the academic staff of a university shall have the freedom, within the law, in his or her teaching, research and any other activities either in or outside the university, to question and test received wisdom, to put forward new ideas and to state controversial or unpopular opinions and shall not be disadvantaged, or subject to less favourable treatment by the university, for the exercise of that freedom" (Government of Ireland 1997). Within this definition of academic freedom, I recognize that writing this provocative book is both my privilege and my duty.

As momentum grows around the world for transformative change to reverse growing economic inequities and reduce climate vulnerabilities (Gahman et al. 2022), academic institutions represent critically important social infrastructure. In addition to universities being places for imagination, struggle, and resistance (Aziz and Salim 2020), the structure and funding of higher education systems has deterministic power in shaping the future. The vision of climate justice universities recognizes this power and acknowledges that higher education has a critically important role in societal transformation. To leverage this power, universities themselves must undergo a fundamental transformation in how they are structured and funded. Intense self-reflection and transparency within higher education are necessary to collectively confront the possibility that some of the knowledge produced and disseminated in universities reinforces a dominant socioeconomic system that is failing most people and devastating the nonhuman world.

Among climate justice advocates and youth activists, the phrase "the future is now" is increasingly used to characterize the immediate urgency for making change; this phrase highlights that decisions made in the present moment are shaping the future (Keller and Heri 2022). While futurists explore different possibilities for what the future might hold (Alexander

2023), the future is in fact more predictable now than it has ever been. We know the future will include more frequent and more intense climate disruptions. We know that sea level will continue to rise and mass species extinctions will accelerate; in addition to growing volatility, we know the many impacts of climate change will continue to get worse (McKibben 2023). Our present collective actions and priorities will influence how we respond to these changes, rendering some aspects of the future inevitable and undeniable.

Education futurist Keri Facer calls for a radical diversification of the space between critique and desire—going beyond complaints and critiques about higher education (like Bill Readings's 1996 book *The University in Ruins*) or pie-in-the-sky imaginations of what universities could be (Facer 2022). Here, in this book, I try to span this divide by offering a bit of both and contributing to the middle road—beyond the incremental reform identified by Sharon Stein (2022). Facer points out that there is a struggle between the colonization of the future and the attempt to keep open the possibility of alternative futures (2022).

The ADAPT-ing acronym introduced in chapter 1 from Dr. Jalonne White-Newsome, an environmental justice analyst serving as the senior director for environmental justice in the Biden-Harris administration's White House Council for Environmental Quality, defines a valuable climate justice framework to guide and orient the priorities of universities. As mentioned, ADAPT-ing stands for Acknowledging the harm; Demanding accountability; Addressing racism, power, and privilege; Prioritizing equity; Transforming systems (White-Newsome 2021). For the higher education sector, these priorities are all relevant and implementable principles, but it is clear that powerful forces are actively resisting each of these.

The first step of *acknowledging harm* is particularly challenging because it is counter to the positive branding, marketing, and communication that has become central to the daily operations of many universities. While some universities have retained a collegial, distributed governance model with distributed and participatory management that relies on a culture of *accountability* (see description of the governance structure of Trinity College Dublin in chapter 5), the corporatized governance of many universities intentionally deflects and resists efforts toward transparency and community accountability. Although many universities make ambitious declarations about their commitments to *addressing racism, power, and privilege*, the lack of transformative and structural change in how universities are organized has constrained impact in this area. While few universities or university leaders will publicly admit to resisting efforts to *promote equity*, the practices and priorities within many universities continue to exacerbate inequities and disparities. The final dimension of White-Newsome's approach to climate justice requires universities to *transform systems*. This is arguably the most basic, yet the most challenging, because although systems thinking has become popular in some contemporary university programs, system transformation is not widely accepted as a mission or purpose of higher education. Although the cooperative Mondragon University in the Basque Country of Spain explicitly focuses on societal transformation (as mentioned in chapter 6), most universities around the world have not yet been so bold.

Changing Direction with Hope, Humility, and Care

The creative exploration of climate justice universities in this book does not prescribe a specific path for specific individuals or institutions. Instead, I have tried to encourage us to stop, reflect, and consider ways that we can slow down, turn around, and change direction. Although there is urgency for climate justice, we must be careful that our sense of urgency does not rush us deeper into crisis. An indigenous climate justice activist spoke with wisdom about this temporal risk at a Climate Community Collaborative event hosted at the Massachusetts Institute of Technology in October 2023 when she said, "It is too late to rush—we must move diligently with extreme care."

In addition to changing direction with care, this work of reimagining universities for climate justice requires both hope and humility. Hope has many meanings and interpretations; it is both a verb and a noun. Recognizing the collective suffering, devastation, and grieving of humanity in this era of polycrisis, hope has been defined as both a discipline and a practice (Hayes and Kaba 2023). Mariame Kaba, a US-based organizer, educator, and activist working to abolish the prison industrial complex, describes hope as a discipline to acknowledge that hope is not a warm and fuzzy feeling of optimism; rather hope takes focus and hard work to get up every day and keep in the struggle.

To change course, to reimagine the regenerative potential role of higher education in society, humility is also essential. Although arrogance and defensiveness are characteristics that are often rewarded and encouraged in universities, both individual humility and institutional humility are needed to leverage the transformative power of higher education to shape a hopeful future during this time of worsening polycrisis.

Establishing a culture of care and accountability in universities requires intentionally resisting dehumanizing frameworks that promote complacency to human suffering and disconnect us from each other and to the regenerative power of the earth's systems. Rather than encouraging disconnection from the nonhuman world, there is an urgent need for higher education institutions to reconceptualize nature not as a resource for extraction but more as an elder relative whose wisdom we can all learn from. If universities humbly recommit themselves to become learning institutions—open to learning from nature and committed to learning from people and communities who may not be on an academic path—a change in direction toward transformative impact will emerge.

For my own trajectory, I am also changing direction. I have moved from the United States back to Ireland, the country where I was born and where I spent the first eight years of my life. In my new position, I am Professor of Climate Justice at the National University of Ireland Maynooth in County Kildare, outside Dublin. At Maynooth University, a higher education institution whose official purpose is "to imagine and create better futures for all," I am collaborating in an academic community with many inspiring colleagues committed to centering justice and global solidarity. My work focuses on co-creating knowledge for a just transition and moving collectively toward the paradigm shift of climate justice universities. In my new position, I am also engaging with the network of academics and activists from all around Ireland who met in Galway to collectively implement a transformative vision aligned with the vision of climate justice universities. We have established a Climate Justice Universities Union to organize our collective power to advocate for universities to become community-engaged, critical infrastructure for implementing ambitious, transformative change toward climate justice. We are in coalition and collaboration with multiple community-based climate justice organizations including Ireland's Feminist Communities for Climate Justice network and the climate cooperative within Dublin's oldest football club (the member-owned Bohemian Football Club, which has a climate justice officer and advocates for climate action for the people by the people). As the need for systemic transformative social change continues to grow, I invite others around the world to join a growing global movement for climate justice universities.

REFERENCES

AAUP. 2023. "Collective Bargaining." American Association of University Professors. Web-based resource. https://www.aaup.org/programs/collective-bargaining.

Abatayo, Anna Lou, Valentina Bosetti, Marco Casari, Riccardo Ghidoni, and Massimo Tavoni. 2020. "Solar Geoengineering May Lead to Excessive Cooling and High Strategic Uncertainty." *Proceedings of the National Academy of Sciences,* 117: 13393–98.

Abrica, Elvira J, Deryl Hatch-Tocaimaza, and Cecilia Rios-Aguilar. 2021. "On the Impossibilities of Advancing Racial Justice in Higher Education Research through Reliance on the Campus Climate Heuristic." *Journal of Diversity in Higher Education,* 16(2): 144–56. https://doi.org/10.1037/dhe0000323.

Ackerman, Frank, and Elizabeth Stanton. 2008. "Climate Change and the US Economy: The Costs of Inaction." https://frankackerman.com/publications/climatechange/Climate_Change_US_Economy.pdf.

Adefarakan, Temitope. 2018. "Integrating Body, Mind, and Spirit through the Yoruba Ori: Critical Contributions to a Decolonizing Pedagogy." In Sheila Batacharya and Yuk-Lin Renita Wong (Eds.), *Sharing Breath: Embodied Learning and Decolonization.* Athabasca University Press. https://doi.org/10.15215/aupress/9781771991919.01.

Ahamed, Sonya, Gillian L. Galford, Bindu Panikkar, Donna Rizzo, and Jennie C. Stephens. 2024. "Carbon Collusion: Cooperation, Competition, and Climate Obstruction in the Global Oil and Gas Extraction Network." *Energy Policy,* 190: 114103. https://doi.org/10.1016/j.enpol.2024.114103.

Aldrich, Daniel P. 2012. "Social, Not Physical, Infrastructure: The Critical Role of Civil Society after the 1923 Tokyo Earthquake." *Disasters,* 36: 398–419.

Alexander, Bryan. 2023. *Universities on Fire: Higher Education in Climate Crisis.* Johns Hopkins University Press.

Ali, Saadia. 2021. *Higher Education as Critical Infrastructure.* https://globalresilience.northeastern.edu/higher-education-as-critical-infrastructure/.

Allen, D. 2023. "We Hit Rock Bottom on Civics Education. Can We Turn It

Around?" *Washington Post*, https://www.washingtonpost.com/opinions/2023/2005/2023/civics-education-democracy-danielle-allen/.

Almond, Douglas, Xinming Du, and Anna Papp. 2022. "Favourability towards Natural Gas Relates to Funding Source of University Energy Centres." *Nature Climate Change*, 12: 1122–28.

Alvares, Claude, and Shad Saleem Faruqi (Eds.). 2012. *Decolonising the University: The Emerging Quest for Non-Eurocentric Paradigms*. Penerbit Universiti Sains Malaysia.

American Petroleum Institute. 1998. "Global Climate Science Communications—Action Plan." Accessed June 1, 2023. https://perma.cc/LJ33-LSEB.

Anderson, Kevin, and Glen Peters. 2016. "The Trouble with Negative Emissions." *Science*, 354: 182.

Annie, Freitas. 2017. "Beyond Acceptance: Serving the Needs of Transgender Students at Women's Colleges." *Humboldt Journal of Social Relations*, 39: 294–314.

Aoun, Joseph E. 2024. "How Higher Ed Can Adapt to the Challenges of AI." *Chronicle of Higher Education*. July 1. https://www.chronicle.com/article/how-higher-ed-can-adapt-to-the-challenges-of-ai.

Arnstein, S. R. 1969. "A Ladder of Citizen Participation." *Journal of the American Planning Association*, 35(4): 216–24.

Aronoff, Kate, Alyssa Battistoni, Daniel Aldana Cohen, and Thea Riofrancos. 2019. *A Planet to Win: Why We Need a Green New Deal*. Verso.

Arora-Jonsson, Seema. 2023. "The Sustainable Development Goals: A Universalist Promise for the Future," *Futures*, 146: 103087.

Arun, Banerji. 1996. "A Long Struggle to Escape from Old Ideas." *Economic and Political Weekly*, 31: 1064–67.

Avelino, F. 2017. "Power in Sustainability Transitions: Analysing Power and (Dis)Empowerment in Transformative Change towards Sustainability." *Environmental Policy and Governance*, 27(6): 505–20. https://doi.org/10.1002/eet.1777.

Avelino, F. 2021. "Theories of Power and Social Change. Power Contestations and Their Implications for Research on Social Change and Innovation." *Journal of Political Power*, 14(3): 1–24. https://doi.org/10.1080/2158379X.2021.1875307.

Avelino, Flor, and Jan Rotmans. 2009. "Power in Transition: An Interdisciplinary Framework to Study Power in Relation to Structural Change." *European Journal of Social Theory*, 12(4): 543–69. https://doi.org/10.1177/1368431009349830.

Bacow, Lawrence. 2022. "Harvard and the Legacy of Slavery." Harvard University. https://www.harvard.edu/president/news/2022/harvard-and-the-legacy-of-slavery/.

Baldwin, Davarian. 2021. *In the Shadow of the Ivory Tower: How Universities Are Plundering Our Cities*. Bold Type Books.
Baldwin, Davarian. 2022. "Building a New Framework of Values for the University." Interview by Jennifer Mittelstadt. *Academe, 108*(4): 20–23.
Banks, Jasmine. 2023. "UnKoching: No Masters, No [Corporate] Gods." *OnlySky*. https://onlysky.media/jbanks/unkoching-no-masters-no-corporate-gods/.
Barrett, Sam. 2013. "The Necessity of a Multiscalar Analysis of Climate Justice." *Progress in Human Geography, 37*: 215–33.
Barry, John. 2011. "Knowledge as Power, Knowledge as Capital: A Political Economy Critique of Modern 'Academic Capitalism.'" *Irish Review, 43*.
Baskaran, Priya. 2015. "Introduction to Worker Cooperatives and Their Role in the Changing Economy." *Journal of Affordable Housing and Community Development Law, 24*: 355–81.
Battle, Collette Pichon. 2020. "An Offering from the Bayou." In Ayana Elizabeth Johnson and Katherine K. Wilkinson (Eds.), *All We Can Save* (329–33). One World.
Benjamin, Ruha. 2023. University of Southern California Commencement Speech. May 2023. https://www.youtube.com/watch?v=rUmmgJZDBfQ.
Benjamin, Ruha. 2024. *Imagination: A Manifesto*. Norton.
BER. 2022. "The Financialization of Education." *Berkeley Economic Review*. https://econreview.berkeley.edu/the-financialization-of-education/.
Bergart, Ann, Jennifer Currin-McCulloch, Kristina Lind, Namoonga B. Chilwalo, Donna Louise Guy, Neil Hall, Diana Kelly, et al. 2023. "We Need Mutual Aid Too: Group Work Instructors Helping Each Other Navigate Online Teaching." *Social Work with Groups, 46*: 21–35.
Bergh, Jeroen C., J. M. van den, Albert Faber, Annemarth M. Idenburg, and Frans H. Oosterhuis. 2007. "Evolutionary Economics and Environmental Policy: Survival of the Greenest." In Geoffrey M. Hodgson. (Ed.), *New Horizons in Institutional and Evolutionary Economics*. Edward Elgar.
Bergmann, Melanie, France Collard, Joan Fabres, Geir W. Gabrielsen, Jennifer F. Provencher, Chelsea M. Rochman, Erik van Sebille, and Mine B. Tekman. 2022. "Plastic Pollution in the Arctic." *Nature Reviews Earth and Environment, 3*: 323–37.
Beyond the Academy. 2022. *Guidebook for the Engaged University: Best Practices for Reforming Systems of Reward, Fostering Engaged Leadership, and Promoting Action Oriented Scholarship*. http://beyondtheacademynetwork.org/guidebook.
Bhambra, Gurminder K., Dalia Gebrial, and Kerem Nişancıoğlu. 2018. "Decolonising the University?" In Gurminder K. Bhambra, Dalia Gebrial, and Kerem Nişancıoğlu (Eds.), *Decolonising the University*. Pluto Press.

Bleemer, Zachary. 2023. "Affirmative Action and Its Race-Neutral Alternatives." *Journal of Public Economics*, 220: 104839.

Blettler, Martín C. M., Elie Abrial, Farhan R. Khan, Nuket Sivri, and Luis A. Espinola. 2018. "Freshwater Plastic Pollution: Recognizing Research Biases and Identifying Knowledge Gaps." *Water Research*, 143: 416–24.

Boggs, Abigail, and Nick Mitchell. 2018. "Critical University Studies and the Crisis Consensus." *Feminist Studies*, 44: 432–63.

Bondestam, Fredrik, and Maja Lundqvist. 2020. "Sexual Harassment in Higher Education: A Systematic Review." *European Journal of Higher Education*, 10: 397–419.

Boyle, Alaina D., and Jennie C. Stephens. 2022. "Higher Education Needs a New Mission: How about Climate Justice?" *Boston Globe*, September 4.

Boyle, M.-E., L.Ross, and Jennie C. Stephens. 2011. "Who Has a Stake? How Stakeholder Processes Influence Partnership Sustainability." *Gateways: International Journal of Community Research and Engagement*, 4: 100–118.

Bracey, E. N. 2017. "The Significance of Historically Black Colleges and Universities (HBCUs) in the 21st Century: Will Such Institutions of Higher Learning Survive?" *American Journal of Economics and Sociology*, 76(3): 670–96.

Brownill, S., and Q Bradley. (Eds.). 2017. *Localism and Neighbourhood Planning: Power to the People?* Bristol University Press.

Brown University. 2022. "Updates to Brown's Business Ethics Policies and Practices." Brown University. Accessed June 19, 2023. https://president.brown.edu/president/updates-browns-business-ethics-policies-and-practices.

Brulle, Robert J., and Riley E. Dunlap. 2023. "A Sociological View of the Effort to Obstruct Action on Climate Change," *Footnotes*, 9. https://www.asanet.org/footnotes-article/sociological-view-effort-obstruct-action-climate-change/.

Bruni, Frank. 2023. "There's Only One College Rankings List That Matters." *New York Times*, March 27.

Brunswick Group. 2017. "Advocacy Campaign Gas and Methane." Accessed May 19, 2023. https://perma.cc/78AL-KSXQ.

Bula, Frances. 2017. "Universities as Real-Estate Developers." *University Affairs*. https://www.universityaffairs.ca/features/feature-article/universities-real-estate-developers/.

Bull, Anna. 2021. "Will Academia Ever Have Its #MeToo Moment?" *Al Jazeera*. https://www.aljazeera.com/opinions/2021/10/28/will-academia-ever-have-its-metoo-moment.

Bullard, Robert D. 1993. *Confronting Environmental Racism: Voices from the Grassroots*. South End Press.

Bullard, Robert D., and Glenn S. Johnson. 2000. "Environmental Justice Grassroots Activism and Its Impact on Public Policy Decision Making." *Journal of Social Issues,* 56: 555–78.

Burch, Audra D. S. 2023. "A New Frontier in Reparations: Seeking the Return of Lost Family Land." *New York Times,* June 8.

Burgen, Stephen. 2022. "Barcelona Students to Take Mandatory Climate Crisis Module from 2024." *Guardian,* November 12.

Burke, Matthew J., and Jennie C. Stephens. 2017. "Energy Democracy: Goals and Policy Instruments for Sociotechnical Transitions." *Energy Research and Social Science,* 33: 35–48.

Byrd, W. Carson. 2021. *Behind the Diversity Numbers: Achieving Racial Equity on Campus.* Harvard Education Press.

Canavan, Gerry. 2014. " 'There's Nothing New / Under the Sun, / But There Are New Suns': Recovering Octavia E. Butler's Lost Parables." *LA Review of Books,* June 9. https://lareviewofbooks.org/article/theres-nothing-new-sun-new-suns-recovering-octavia-e-butlers-lost-parables/.

Cappelli, Federica, Valeria Costantini, and Davide Consoli. 2021. "The Trap of Climate Change–Induced 'Natural' Disasters and Inequality." *Global Environmental Change,* 70: 102329.

Carastathis, Anna. 2016. "Interlocking Systems of Oppression." In Nelson M. Rodriguez, Wayne J. Martino, Jennifer C. Ingrey, and Edward Brockenbrough (Eds.), *Critical Concepts in Queer Studies and Education: An International Guide for the Twenty-First Century.* Palgrave Macmillan.

Carlson, Colin J., Rita Colwell, Mohammad Sharif Hossain, Mohammed Mofizur Rahman, Alan Robock, Sadie J. Ryan, Mohammad Shafiul Alam, and Christopher H. Trisos. 2022. "Solar Geoengineering Could Redistribute Malaria Risk in Developing Countries." *Nature Communications,* 13: 2150.

Carlton, Genevieve. 2023. "A History of Women in Higher Education." *Best Colleges.* https://www.bestcolleges.com/news/analysis/2021/03/21/history-women-higher-education/.

Carroll, William K. 2021. *Regime of Obstruction: How Corporate Power Blocks Energy Democracy.* Athabasca University Press.

Carroll, William K., N. Graham, and Y. Zunker. 2018. "Carbon Capital and Corporate Influence: Mapping Elite Networks of Corporations, Universities and Research Institutes." In J. Brownlee, C. Hurl, and K. Walby (Eds.), *Minding the Public's Business: Critical Perspectives on Corporatization in Canada.* Between the Lines.

Casale, R. 2019. *Civid Revolution.* Metador.

Casey, Zachary A. (2017). *A Pedagogy of Anticapitalist Antiracism.* SUNY Press.

CBGLC. 2023. "Fair Trade Learning." Community-Based Global Learning Collaborative. Accessed June 20, 2023. https://www.cbglcollab.org/ftl.

Chankseliani, Maia, and Tristan McCowan. 2021. "Higher Education and the Sustainable Development Goals." *Higher Education*, 81: 1–8.

Choudry, Aziz, and Salim Vally, Eds. 2020. *The University and Social Justice: Struggles across the Globe*. Pluto Press.

Chronicle of Higher Education. 2022. "How Much Are Private-College Presidents Paid?" *Chronicle of Higher Education*, February 15. https://www.chronicle.com/article/president-pay-private-colleges.

Chrysopoulou, Anna. 2020. "The Vision of a Well-Being Economy." *Stanford Social Innovation Review*. https://ssir.org/articles/entry/the_vision_of_a_well_being_economy#.

Chung, Jane. 2022. "Besiege the Ivy League." *American Prospect*. https://prospect.org/education/besiege-the-ivy-league/.

Clarence-Smith, Suryamayi Aswini. 2015. "Auroville: A Practical Experiment in Utopian Society." Undergraduate thesis. University of California Berkeley. https://www.academia.edu/24723165/Auroville_A_Practical_Experiment_in_Utopian_Society.

Clarence-Smith, Suryamayi. 2023. *Prefiguring Utopia: The Auroville Experiment*. Bristol University Press.

Clarence-Smith, Suryamayi, and Lara Monticelli. 2022. "Flexible Institutionalisation in Auroville: A Prefigurative Alternative to Development." *Sustainability Science*, 17: 1171–82.

Clem, R. L., and D. Schiller. 2016. "New Learning and Unlearning: Strangers or Accomplices in Threat Memory Attenuation?" *Trends in Neurosciences*, 39: 340–51.

Cohen, P. 2023. "Failures of Globalization Shatter Long-Held Beliefs: War and Pandemic Highlight Shortcomings of the Free-Market Consensus." *New York Times*, June 18.

Colbert, Max. 2023. "Revealed: The Charles Koch Foundation's Staggering Donations to UK's Top Universities." *Byline Times*.

Collins, Katie. 2021. "Women Are Climate Leaders, but They Struggled to Be Heard at COP26." CNET. https://www.cnet.com/news/politics/women-are-our-climate-leaders-but-at-cop26-they-struggled-to-be-heard/.

Connell, Raewyn. 2019. *The Good University: What Universities Actually Do and Why It's Time for Radical Change*. Zed Books.

Contreras, Joseph. 2023. "DeSantis Ramps Up 'War on Woke' with New Attacks on Florida Higher Education." *Guardian*, February 5. https://www.theguardian.com/us-news/2023/feb/05/ron-desantis-war-on-woke-florida-higher-education-new-college.

Copenhagen Business School. 2023. "Minor in Transformative and Sustainable Economies." https://studieordninger.cbs.dk/2022/minor/1011.

Corderoy, Jenna, Billy Stockwell, Martin Williams, and Finlay Johnston. 2023. "Big Oil Given Direct Influence over University Courses." *openDemocracy*. https://www.opendemocracy.net/en/dark-money-investigations/shell-bp-equinor-totalenergies-oxford-cambridge-bristol-ucl-big-oil-gas-fossil-fuels-158m/.

Cortina, Regina, Linda Martin Alcoff, Abadio Green Stocel, and Gustavo Esteva. 2019. "Decolonial Trends in Higher Education: Voices from Latin America." *Compare: A Journal of Comparative and International Education*, 49: 489–506.

Cottom, Tressie McMillan. 2017. *Lower Ed: The Troubling Rise of For-Profit Colleges in the New Economy*. The New Press.

Cownie, Fiona. 2021. "Could Universities Be Guilty of 'Greenwashing'?" *WONKHE*. https://wonkhe.com/blogs/could-universities-be-guilty-of-greenwashing/.

Cox, Ronald W. 2013. "The Corporatization of Higher Education." *Class, Race and Corporate Power*, 1.

Cronin, Michael. 2019. *Irish and Ecology/An Ghaeilge agus an Éiceolaíocht*. Foilseacháin Ábhair Spioradálta.

CSSN. 2021. "The Structure of Obstruction: Understanding Opposition to Climate Change Action in the United States." In *CSSN Primer 2021:1*. Climate Social Science Network.

Cupples, Julie, and R. Grosfoguel (Eds.). 2019. *Unsettling Eurocentrism in the Westernized University*. Routledge.

Daggett, Cara. 2018. "Petro-masculinity: Fossil Fuels and Authoritarian Desire." *Millennium* 47(1): 25–44.

Daly, Herman. 2015. *Essays Against Growthism*. World Economics Association Books.

Dark Matter Labs. 2023. "About Dark Matter Labs." https://darkmatterlabs.org/About.

Data for Progress. 2023. "Accountable Allies: The Undue Influence of Fossil Fuel Money in Academia."

Davidson, Debra J. 2019. "Exnovating for a Renewable Energy Transition." *Nature Energy*, 4: 254–56.

Davies, Emily. 1866. *The Higher Education of Women*. Alexander Strahan.

de Freitas Netto, Sebastião Vieira, Marcos Felipe Falcão Sobral, Ana Regina Bezerra Ribeiro, and Gleibson Robert da Luz Soares. 2020. "Concepts and Forms of Greenwashing: A Systematic Review." *Environmental Sciences Europe*, 32: 19.

de Onís, C. M. 2018. "Energy Colonialism Powers the Ongoing Unnatural Disaster in Puerto Rico." *Frontiers in Communication, 3*.

de Onís, C. M. 2021. *Energy Islands: Metaphors of Power, Extractivism, and Justice in Puerto Rico*. University of California Press.

DEAL. 2023. "DEAL (Doughnut Economics Action Lab)." https://doughnuteconomics.org/.

Denton, Fatma. 2002. "Climate Change Vulnerability, Impacts, and Adaptation: Why Does Gender Matter." *Gender and Development, 10*: 10–20.

Desai, Chandni. 2024. "The War in Gaza Is Wiping Out Palestine's Education and Knowledge Systems." *Conversation*. https://theconversation.com/the-war-in-gaza-is-wiping-out-palestines-education-and-knowledge-systems-222055#.

de Sousa Santos, Boaventura (Ed.). 2008. *Another Knowledge Is Possible: Beyond Northern Epistemologies*. Verso Books.

Dixson-Declève, Sandrine, Owen Owen Gaffney, Jayati Jayati Ghosh, Jørgen Randers, Johan Rockström, and Per Espen Stoknes. 2022. *Earth for All: A Survival Guide for Humanity*. A Report to the Club of Rome.

Douglass, John A. (Ed.) 2021. *Neo-Nationalism and Universities: Populists, Autocrats, and the Future of Higher Education*. Johns Hopkins University Press.

Dunlap, Riley E., and Robert J. Brulle. 2020. "Sources and Amplifiers of Climate Change Denial." In David C. Holmes and Lucy M. Richardson (Eds.), *Research Handbook on Communicating Climate Change*. Edward Elgar Publishing.

Dupas, Pascaline, Alicia Sasser Modestino, Muriel Niederle, Justin Wolfers, and the Seminar Dynamics Collective. 2021. "Gender and the Dynamics of Economics Seminars." National Bureau of Economic Research.

Dutta, Urmitapa 2018. "Decolonizing 'Community' in Community Psychology." *American Journal of Community Psychology, 62*(3–4): 272–82. https://doi.org/10.1002/ajcp.12281.

Eaton, Charlie. 2022. *Bankers in the Ivory Tower: The Troubling Rise of Financiers in US Higher Education*. University of Chicago Press.

Eaton, Charlie, Jacob Habinek, Adam Goldstein, Cyrus Dioun, Daniela García Santibáñez Godoy, and Robert Osley-Thomas. 2016. "The Financialization of US Higher Education." *Socio-Economic Review, 14*: 507–35.

Ecoversities Alliance. 2020. *The Future Is Now: Ecoversities Alliance Catalog of Radical Pedagogies*. https://ecoversities.org/wp-content/uploads/2020/11/Ecoversities-Coffee-tablefinal2.0.pdf.

Educating for American Democracy. 2021. *Educating for American Democracy: Excellence in History and Civics for All Learners*. https://www.educatingforamericandemocracy.org/.

Eells, Walter Crosby. 1934. "Criticisms of Higher Education." *Journal of Higher Education*, 5: 187–89.
Ekberg, Kristoffer, Bernhard Forchtner, Martin Hultman, and Kirsti M. Jylhä. 2022. *Climate Obstruction: How Denial, Delay and Inaction Are Heating the Planet*. Routledge.
Enck, Judith, and Jan Dell. 2022. "Plastic Recycling Doesn't Work and Will Never Work." *Atlantic*.
Engelen, Ewald, Rodrigo Fernandez, and Reijer Hendrikse. 2014. "How Finance Penetrates its Other: A Cautionary Tale on the Financialization of a Dutch University." *Antipode, 46*: 1072–91.
Engle, Jayne, Julian Agyeman, and Tanya Chung-Tiam-Fook. (Eds.). 2022. *Sacred Civics: Building Seven Generation Cities*. Routledge.
ERC. 2023. "Pathways towards Post-Growth Deals." European Research Council. Accessed June 1, 2023. https://erc.europa.eu/news-events/news/erc-synergy-grants-2022-project-highlights.
Erickson, Jon D. 2022. *The Progress Illusion*. Island Press.
ESSA. 2021. "The Gender Gap in Universities and Colleges in Sub-Saharan Africa." https://essa-africa.org/node/1421: Education Sub Saharan Africa.
evans, tina lynn. 2012. *Occupy Education: Living and Learning Sustainability*. Peter Lang.
Eynaud, Philippe, and Genauto Carvalho de Franca Filho. 2023. *Solidarity and Organization toward New Avenues for Management*. Palgrave MacMillan. https://link.springer.com/book/10.1007/978-3-031-27568-5.
Fabbri, A., A. Lai, Q. Grundy, and L. A. Bero. 2018. "The Influence of Industry Sponsorship on the Research Agenda: A Scoping Review." *American Journal of Public Health, 108*: e9–e16.
Faber, Daniel, Jennie C. Stephens, Victor Wallis, Roger Gottlieb, Charles Levenstein, Patrick CoatarPeter, and Boston Editorial Group of CNS. 2017. "Trump's Electoral Triumph: Class, Race, Gender, and the Hegemony of the Polluter-Industrial Complex." *Capitalism Nature Socialism, 28*: 1–15.
Facer, Keri. 2019. "Learning to Live with a Lively Planet: The Renewal of the University's Mission in the Era of Climate Change." Lecture at Uppsala University. http://media.medfarm.uu.se/play/video/9204.
Facer, Keri. 2022. "Imagination and the Future University: Between Critique and Desire." *Critical Times, 5*(1): 202–216. https://doi.org/10.1215/26410478-9536559.
Faculty for a Future. 2023. "Research Toolkit." Accessed June 20, 2023. https://facultyforafuture.org/research.
Fanning, Andrew L., and Jason Hickel. 2023. "Compensation for Atmospheric Appropriation." *Nature Sustainability*.
Farah, Paolo D., and Kirk W. Junker. 2021. *Globalization, Environmental Law*

and Sustainable Development in the Global South: Challenges for Implementation. Routledge.

Favretti, Maggie. 2023. *Learning in the Age of Climate Disasters: Teacher and Student Empowerment beyond Futurphobia.* Routledge.

Fazackerley, Anna. 2023. "'Too Greedy': Mass Walkout at Global Science Journal over 'Unethical' Fees." *Guardian,* May 7. https://www.theguardian.com/science/2023/may/07/too-greedy-mass-walkout-at-global-science-journal-over-unethical-fees?CMP=share_btn_tw.

Fernandez, Glenn, Tong Thi My Thi, and Rajib Shaw. 2014. "Climate Change Education: Recent Trends and Future Prospects." In Rajib Shaw and Yukihiko Oikawa *(Eds.), Education for Sustainable Development and Disaster Risk Reduction.* Springer.

Finneman, Teri, Meg Heckman, and Pamela E. Walck. 2022. "Reimagining Journalistic Roles: How Student Journalists Are Taking On the U.S. News Desert Crisis." *Journalism Studies,* 23(3): 338–55. https://doi.org/10.1080/1461670X.2021.2023323.

Fioramonti, Lorenzo. 2016. *Well-Being Economy: A Scenario for a Post-Growth Horizontal Governance System.* Next System Project. https://thenextsystem.org/sites/default/files/2017-08/LorenzoFioramonti.pdf.

Fitzgerald, Louise Michelle. 2023. "Tracing the Development of Anti-Fossil Fuel Norms: Insights from the Republic of Ireland." *Climate Policy*: 1–14.

Fitzpatrick, Kathleen. 2019. *Generous Thinking: A Radical Approach to Saving the University.* Johns Hopkins University Press.

Fleck, Anna. 2022. "Gender Gap Persists in U.S. Universities' Leadership." *World Economic Forum.* https://www.weforum.org/agenda/2022/05/gender-gap-us-universities-leadership.

Fontana, Guiseppe, and Malcolm Sawyer. 2016. "Towards Post-Keynesian Ecological Macroeconomics. Ecological Economics." *Ecological Economics,* 121: 186–95.

Foroohar, Rana. 2016. "How the Financing of Colleges May Lead to Disaster!" *New York Review,* October 13. https://www-nybooks-com.ezproxy.neu.edu/articles/2016/10/13/how-the-financing-of-colleges-may-lead-to-disaster/.

Forrest, Andrew, Luca Giacovazzi, Sarah Dunlop, Julia Reisser, David Tickler, Alan Jamieson, and Jessica J. Meeuwig. 2019. "Eliminating Plastic Pollution: How a Voluntary Contribution from Industry Will Drive the Circular Plastics Economy." *Frontiers in Marine Science,* 6.

Foster, John and J. Stanley Metcalfe. 2001. *Frontiers of Evolutionary Economics: Competition, Self-Organization, and Innovation Policy.* Edward Elgar.

François, Martin, Sybille Mertens de Wilmars, and Kevin Maréchal. 2023. "Unlocking the Potential of Income and Wealth Caps in Post-Growth

Transformation: A Framework for Improving Policy Design." *Ecological Economics*, 208: 107788.

Franta, Benjamin. 2021. "Early Oil Industry Disinformation on Global Warming." *Environmental Politics*, 30: 663–68.

Franta, Benjamin, and Geoffrey Supran. 2017. "The Fossil Fuel Industry's Invisible Colonization of Academia." *Guardian*, March 13.

Friere, Paulo. 1970. *Pedagogies of the Oppressed*. Continuum International Publishing Group. New York.

Fullerton, John. 2015. "Regenerative Economies for a Regenerative Civilization." *Kosmos: Journal for Global Transformation*. https://www.kosmosjournal.org/article/regenerative-economies-for-a-regenerative-civilization/.

Gahman, Levi, Nasha Mohamed, Filiberto Penados, Johannah-Rae Reyes, Atiyah Mohamed, and Shelda-Jane Smith. 2022. *A Beginner's Guide to Building Better Worlds—Ideas and Inspiration from the Zapatistas*. Bristol University Press.

Gallagher, Patrick. 1939. *My Story: Paddy the Cope—Patrick Gallagher's Autobiography*. Templecrone Press.

Galligan, Yvonne. 2022. "Gender Inequality in Higher Education." *European Association for International Education*. https://www.eaie.org/blog/gender-inequality-ireland.html.

Gallup/Knight Foundation. 2020. *American Views 2020: Trust, Media and Democracy: A Deepening Divide*. https://knightfoundation.org/wp-content/uploads/2020/08/American-Views-2020-Trust-Media-and-Democracy.pdf.

Gearino, Dan. 2023. "Students and Faculty at Ohio State Respond to a Bill that Would Restrict College Discussions of Climate of Climate Policies." *Inside Climate News*. https://insideclimatenews.org/news/31052023/ohio-state-college-climate-bill/.

Geels, Frank W. 2014. "Regime Resistance against Low-Carbon Transitions: Introducing Politics and Power into the Multi-Level Perspective." *Theory, Culture and Society*, 31: 21–40.

Ghosh, Amitav. 2017. *The Great Derangement: Climate Change and the Unthinkable*. University of Chicago Press.

Gibson-Graham, J. K., Jenny Cameron, and Stephen Healy. 2013. *Take Back the Economy: An Ethical Guide for Transforming our Communities*. University of Minnesota Press.

Gluckman, Nell. 2017. "'A Complete Culture of Sexualization': 1,600 Stories of Harassment in Higher Ed." *Chronicle of Higher Education*. https://www.chronicle.com/article/a-complete-culture-of-sexualization-1-600-stories-of-harassment-in-higher-ed/.

Goddard, John, Ellen Hazelkorn, Louise Kempton, and Paul Vallance. (Eds.). 2016. *The Civic University: The Policy and Leadership Challenges.* Edward Elgar.

Gosha, Ryan. 2022. "Socialization of Losses—A Crisis of Capitalism." https://ryangosha.medium.com/socialization-of-losses-a-crisis-of-capitalism-878fcf347d54.

Government of Ireland. Irish Universities Act. 1997. https://www.irishstatutebook.ie/eli/1997/act/24/enacted/en/html.

Graham, N. 2020. "Fossil Knowledge Networks: Science, Ecology, and the 'Greening' of Carbon Extractive Development." *Studies in Political Economy,* 101: 93–113. https://doi.org/10.1080/07078552.2020.1802831.

Gramsci, Antonio. 2011. *Prison Notebooks.* (Joseph A. Buttigieg, Ed.). Columbia University Press.

Grasso, Marco, and Richard Heede. 2023. "Time to Pay the Piper: Fossil Fuel Companies' Reparations for Climate Damages." *One Earth,* 6: 459–63.

Grauerholz, Elizabeth. 1989. "Sexual Harassment of Women Professors by Students: Exploring the Dynamics Of Power, Authority, and Gender in a University Setting." *Sex Roles,* 21: 789–801.

Gray, Gary, and William K. Carroll. 2018. "Mapping Corporate Influence and Institutional Corruption Inside Canadian Universities." *Critical Criminology* 26: 491–507.

Gray, Hanna Holborn. 2001. "The University in History: 1088 and All That." Remarks presented at the Idea of the University Colloquim, January 17, 2001, https://iotu.uchicago.edu/gray.html.

Green, Fergus. 2018. "The Logic of Fossil Fuel Bans." *Nature Climate Change,* 8: 449–51.

Green, Fergus. 2022. "Fossil Free Zones: A Proposal." *Climate Policy*: 1–7.

Green, Jeremy. 2022. "Comparative Capitalisms in the Anthropocene: A Research Agenda for Green Transition." *New Political Economy*: 1–18.

Grindell, Cheryl, Elizabeth Coates, Liz Croot, and Alicia O'Cathain. 2022. "The Use of Co-production, Co-design and Co-creation to Mobilise Knowledge in the Management of Health Conditions: A Systematic Review." *BMC Health Services Research,* 22: 877.

Gupta, Joyeeta, Diana Liverman, Klaudia Prodani, Paulina Aldunce, Xuemei Bai, Wendy Broadgate, Daniel Ciobanu, et al. 2023. "Earth System Justice Needed to Identify and Live within Earth System Boundaries." *Nature Sustainability,* 6: 630–38.

Guterres, António. 2023. "Secretary-General's Video Message for Press Conference to Launch the Synthesis Report of the Intergovernmental Panel on Climate Change." United Nations. https://www.un.org/sg/en/content/sg/statement/2023-03-20/secretary-generals-video-message-for-press

-conference-launch-the-synthesis-report-of-the-intergovernmental-panel-climate-change.

Guy-Evans, Olivia. 2023. "Malestream: Feminist Critique of Sociology." *Simply Sociology*. https://simplysociology.com/malestream.html.

Hall, Rachel. 2021. "Sexual Harassment Rife in UK Universities, Warns Staff Union." *Guardian*, December 21. https://www.theguardian.com/education/2021/dec/22/sexual-harassment-rife-in-uk-universities-warns-staff-union.

Hall, Shannon. 2023. "A Mental-Health Crisis Is Gripping Science—Toxic Research Culture Is to Blame." *Nature*. https://www.nature.com/articles/d41586-023-01708-4.

Hamann, Julian, and Leopold Ringel. 2023. "The Discursive Resilience of University Rankings." *Higher Education*.

Hamer, Fannie Lou. 1971. "Nobody's Free Until Everybody's Free." Speech Delivered at the founding of the National Women's Political Caucus, Washington, DC, July 10, 1971. In M. P. Brooks, D. W. Houck, M. P. Brooks, and D. W. Houck (Eds.), *The Speeches of Fannie Lou Hamer: To Tell It Like It Is*. University Press of Mississippi, 2010.

Haraway, Donna. 1988. "Situated Knowledges: The Science Question in Feminism and the Privelege of Partial Perspective." *Feminist Studies, 14*: 575–99.

Harlan, Sharon L., David N. Pellow, and J. Timmons Roberts. 2015. "Climate Justice and Inequality." In Riley E. Dunlap and Robert J. Brulle (Eds.), *Climate Change and Society: Sociological Perspectives*. Oxford.

Harris, Douglas N., and Jonathan Mills. 2021. "Optimal College Financial Aid: Theory and Evidence on Free College, Early Commitment, and Merit Aid from an Eight-Year Randomized Trial." Annenberg, Brown University. https://edworkingpapers.com/sites/default/files/ai21-393.pdf.

Harvard. 2022. "Harvard and the Legacy of Slavery." https://legacyofslavery.harvard.edu/report.

Hayden, Anders, and Clay Dasilva. 2022. "The Wellbeing Economy: Possibilities and Limits in Bringing Sufficiency from the Margins into the Mainstream." *Frontiers in Sustainability, 3*.

Hayes, Kelly, and Mariame Kaba. 2023. "Hope Is a Practice and a Discipline: Building a Path to a Counterculture of Care." *Nonprofit Quarterly*. (Winter).

Healy, N., and J. Debski. 2017. "Fossil Fuel Divestment: Implications for the Future of Sustainability Discourse and Action within Higher Education." *Local Environment, 22*: 699–742.

Healy, Noel, Jennie C. Stephens, and Stephanie A. Malin. 2019a. "Embodied Energy Injustices: Unveiling and Politicizing the Transboundary Harms of

Fossil Fuel Extractivism and Fossil Fuel Supply Chains." *Energy Research and Social Science, 48*: 219–34.
Healy, Noel, Jennie C. Stephens, and Stephanie A. Malin. 2019b. "Fossil Fuels Are Bad for Your Health and Harmful in Many Ways Besides Climate Change." *Conversation*, February 7. https://theconversation.com/fossil-fuels-are-bad-for-your-health-and-harmful-in-many-ways-besides-climate-change-107771.
Heller, Nathan. 2023. "The End of the English Major." *New Yorker*.
Helmore, Edward. 2023. "US University Presidents to Testify before Congress over Claims of Antisemitic Protests on Campus." *Guardian*, November 28. https://www.theguardian.com/us-news/2023/nov/28/university-presidents-testify-congress-antisemitism-harvard-mit.
Hernández, Diana, and Stephen Bird. 2010. "Energy Burden and the Need for Integrated Low-Income Housing and Energy Policy." *Poverty Public Policy,* 2: 5–25.
Hernando, Marcos Gonzalez, and Gerry Mitchell. 2023. *Uncomfortably Off: Why the Top 10% of Earners Should Care about Inequality*. Policy Press.
Hickel, Jason. 2021. *Less Is More: How Degrowth Will Save the World*. Penguin.
Hickel, Jason. 2023. "Universal Public Services: The Power of Decommodifying Survival." https://www.jasonhickel.org/blog/2023/3/18/universal-public-services.
Hickel, Jason, Giorgos Kallis, Tim Jackson, Daniel W. O'Neill, Juliet B. Schor, Julia K. Steinberger, Peter A. Victor, and Diana Ürge-Vorsatz. 2022. "Degrowth Can Work—Here's How Science Can Help." *Nature,* 612: 400–403.
Hickman, Caroline, Elizabeth Marks, Panu Pihkala, Susan Clayton, R. Eric Lewandowski, Elouise E. Mayall, Britt Wray, Catriona Mellor, and Lise van Susteren. 2021. "Climate Anxiety in Children and Young People and Their Beliefs about Government Responses to Climate Change: A Global Survey." *Lancet Planetary Health, 5*: e863–e73.
Higgins, Michael D. 2023. "Speech to TASC (Think-Tank for Action on Social Change)." April 23. Aras an Uachtarain, Dublin, Ireland.
Hiltner, Sofia, Emily Eaton, Noel Healy, Andrew Scerri, Jennie C. Stephens, and Geoffrey Supran. 2024. "Fossil Fuel Industry Influence in Academia: A Research Agenda." *WIRES Wiley Interdisciplinary Reviews*.
Hoffman, Andrew J. 2021. "Business Education as If People and the Planet Really Matter." *Strategic Organization, 19*(3): 513–25.
Hoinle, Birgit, Ilke Roose, and Himanshu Shekhar. 2021. "Creating Transdisciplinary Teaching Spaces. Cooperation of Universities and Non-University Partners to Design Higher Education for Regional Sustainable Transition." *Sustainability, 13*(7): 3680.

Honeycutt, N, ST Stevens, and E Kaufmann. 2023. "The Academic Mind in 2022: What Faculty Think about Free Expression and Academic Freedom on Campus." Foundation for Individual Rights and Expression. https://www.thefire.org/research-learn/academic-mind-2022-what-faculty-think-about-free-expression-and-academic-freedom.

hooks, bell. 1994. *Teaching to Transgress: Education as the Practice of Freedom*. Routledge.

hooks, bell. 2003. *Teaching Community: A Pedagogy of Hope*. Routledge.

Horton, Alice A. 2022. "Plastic Pollution: When Do We Know Enough?" *Journal of Hazardous Materials*, 422: 126885.

Howitt, Richard. 2020. "Decolonizing People, Place and Country: Nurturing Resilience across Time and Space." *Sustainability*, 12: 5882.

Hurth, Victoria, and Iain S. Stewart. 2022. "Re-purposing Universities: The Path to Purpose." *Frontiers in Sustainability*, 2. https://doi.org/10.3389/frsus.2021.762271.

Ilies, Remus, Nancy Hauserman, Susan Schwochau, and John Stibal. 2003. "Reported Incidence Rates of Work-Related Sexual Harassment in the United States: Using Meta-Analysis to Explain Reported Rate Disparities." *Personnel Psychology*, 56: 607–31.

IPCC. 2005. "IPCC Special Report on Carbon Dioxide Capture and Storage, Summary for Policymakers." Intergovernmental Panel on Climate Change, Working Group III.

IPCC. 2007. "Climate Change 2007, The Fourth Assessment Synthesis Report." Intergovernmental Panel on Climate Change.

IPCC. 2014. "Synthesis Report, Fifth Assessment." Intergovernmental Panel on Climate Change.

IPCC. 2021. "Climate Change 2021: The Physical Science Basis. Contribution of Working Group I." In Masson-Delmotte V., P. Zhai, A. Pirani, S.L. Connors, C. Péan, S. Berger, N. Caud, et al. (Eds.), *The Sixth Assessment Report of the Intergovernmental Panel on Climate Change*. Cambridge University Press.

IPCC. 2022a. *Climate Change 2022: Impacts, Adaptation, and Vulnerability. Contribution of Working Group II to the Sixth Assessment Report of the Intergovernmental Panel on Climate Change*. Cambridge University Press.

IPCC. 2022b. "Summary for Policymakers." In H.-O. Pörtner, D.C. Roberts, E.S. Poloczanska, K. Mintenbeck, M. Tignor, A. Alegría, M. Craig, S. Langsdorf, S. Löschke, V. Möller and A. Okem (Eds.), *Climate Change 2022: Impacts, Adaptation, and Vulnerability. Contribution of Working Group II to the Sixth Assessment Report of the Intergovernmental Panel on Climate Change*.

IPCC. 2023. "Summary for Policymakers." In *Climate Change 2023: Synthesis*

Report. Contribution of Working Groups I, II and III to the Sixth Assessment Report of the Intergovernmental Panel on Climate Change. Intergovernmental Panel on Climate Change, 1–34, https://doi.org/10.59327/IPCC/AR6-9789291691647.001.

Iroegbu, Austine Ofondu Chinomso, Suprakas Sinha Ray, Vuyelwa Mbarane, João Carlos Bordado, and José Paulo Sardinha. 2021. "Plastic Pollution: A Perspective on Matters Arising: Challenges and Opportunities." *ACS Omega*, 6: 19343–55.

Jackson, Tim. 2022. *Prosperity without Growth: Foundations for the Economy of Tomorrow*. Routledge.

Jain, Manish, and Kalashree Senggupta. 2023. "The Journey to Alivehoods." *Samuhik Pahal*, 3.

Jenner, S., P. Djermester, J. Prügl, C. Kurmeyer, and S. Oertelt-Prigione. 2019. "Prevalence of Sexual Harassment in Academic Medicine." *JAMA Internal Medicine*, 179: 108–11.

Jessop, Bob. 2018. "On Academic Capitalism." *Critical Policy Studies*, 12: 104–9.

Joseph, Tiffany D., and Laura E. Hirshfield. 2023. "Reexamining Racism, Sexism, and Identity Taxation in the Academy." *Ethnic and Racial Studies*, 46: 1101–8.

Kahle, Brewster. 2023. "The US Library System, Once the Best in the World, Faces Death by a Thousand Cuts." *Guardian*, October 9.

Kaidesoja, Tuukka. 2022. "A Theoretical Framework for Explaining the Paradox of University Rankings." *Social Science Information*, 61: 128–53.

Kallis, Giorgos, and Richard B. Norgaard. 2010. "Coevolutionary Ecological Economics." *Ecological Economics*, 69: 690–99.

Kane, Sally, and Jason F. Shogren. 2000. "Linking Adaptation and Mitigation in Climate Change Policy." *Climatic Change*, 45: 75–102.

Keller, Helen, and Corina Heri. 2022. "The Future Is Now: Climate Cases before the ECtHR." *Nordic Journal of Human Rights*, 40(1): 153–74. https://doi.org/10.1080/18918131.2022.2064074.

Kelly, Goergia, and Shaula Massena. 2009. "Mondragon Worker-Cooperatives Decide How to Ride Out a Down-turn." *Yes! Solutions Journalism*. https://www.yesmagazine.org/issue/new-economy/2009/06/06/mondragon-worker-cooperatives-decide-how-to-ride-out-a-downturn.

Kelly, Marjorie. 2023. *Wealth Supremacy: How the Extractive Economy and the Biased Rules of Capitalism Drive Today's Crises*. Berrett-Koehler.

Kelly, Orla, Sam Illingworth, Fabrizio Butera, Vaille Dawson, Peta White, Mindy Blaise, Pim Martens, et al. 2022. "Education in a Warming World: Trends, Opportunities and Pitfalls for Institutes of Higher Education." *Frontiers in Sustainability*, 3.

Kelly, Orla, Peta White, Fabrizio Butera, Sam Illingworth, Pim Martens, Maud

Huynen, Susan Bailey, Geertje Schuitema, and Sian Cowman. 2023. "A Transdisciplinary Model for Teaching and Learning for Sustainability Science in a Rapidly Warming World." *Sustainability Science.*

Kelton, S. 2020. *The Deficit Myth: Modern Monetary Theory and How to Build a Better Economy.* John Murray.

Kendi, Ibram X. 2019. *How to Be an Antiracist.* One World.

Kenner, Dario. 2019. *Carbon Inequality: The Role of the Richest in Climate Change.* Routledge.

Ker, Ian. 2011. "Newman's Idea of a University and Its Relevance for the 21st Century." *Australian eJournal of Theology, 18.*

Kidd, Celeste, and Abeba Birhane. 2023. "How AI Can Distort Human Beliefs." *Science, 380*: 1222–23.

Kim, Anne. 2017. "The Push for College-Endowment Reform." *Atlantic.* https://www.theatlantic.com/education/archive/2017/10/the-bipartisan-push-for-college-endowment-reform/541140/.

Kim, Areil H., and Meimei Xu. 2023. "Dan Schrag, Top Climate Scientist, Faces Allegations of Bullying and Toxicity Spanning Two Decades." *Harvard Crimson.* https://www.thecrimson.com/article/2023/4/28/harvard-climate-professor-schrag-faces-bullying-toxicity-allegations/.

Kimmerer, Robin Wall. 2013. *Braiding Sweetgrass: Indigenous Wisdom, Scientific Knowledge, and the Teachings of Plants.* Milkweed Editions.

Kinol, Alaina, Elijah Miller, Hannah Axtell, Ilana Hirschfeld, Sophie Leggett, Yutong Si, and Jennie C. Stephens. 2023. "Climate Justice in Higher Education: A Proposed Paradigm Shift towards a Transformative Role for Colleges and Universities." *Climatic Change, 176*(2): 15. https://doi.org/10.1007/s10584-023-03486-4.

Kinol, Alaina, Yutong Si, John Kinol, and Jennie C. Stephens. In Press. "Networks of Climate Obstruction: Discourses of Denial and Delay in US Fossil Energy, Plastic, and Agrichemical Industries." *PLOS Climate.*

Klein, Naomi. 2011. "Capitalism vs. the Climate." *The Nation*, November 9.

Klein, Naomi. 2014. *This Changes Everything: Capitalism vs. the Climate.* Simon and Schuster.

Kuhl, Laura, Marla Perez-Lugo, Carlos Arriaga Serrano, Cecelio Ortiz-García, Ryan Ellis, and Jennie C. Stephens. 2024. "Crises, Coloniality, and Energy Transformations in Puerto Rico." In Farhana Sultana (Ed.), *Confronting Climate Coloniality.* Routledge.

Kuhl, Laura, Jennie C. Stephens, Carlos Arriaga Serrano, Marla Perez-Lugo, Cecilio Ortiz Garcia, and Ryan Ellis. 2024. "Fossil Fuel Interest in Puerto Rico: Perceptions of Incumbent Power and Discourses of Delay." *Energy Research and Social Science, 111*: 103467.

Kumar, Bella. 2023. "Accountable Allies: The Undue Influence of Fossil Fuel

Money in Academia." Data for Progress. https://www.dataforprogress.org/memos/accountable-allies-the-undue-influence-of-fossil-fuel-money-in-academia.

Lacey-Barnacle, M., A. Smith, and T. J. Foxon. 2023. "Community Wealth Building in an Age of Just Transitions: Exploring Civil Society Approaches to Net Zero and Future Research Synergies." *Energy Policy, 172*: 113277.

Lachapelle, Paul, Patrick Belmont, Marco Grasso, Roslynn McCann, Dawn H. Gouge, Jerri Husch, Cheryl de Boer, Daniela Molzbichler, and Sarah Klain. 2024. "Academic Capture in the Anthropocene: A Framework to Assess Climate Action in Higher Education." *Climatic Change 177*(3): 40.

Ladd, Anthony E. 2020. "Priming the Well: 'Frackademia' and the Corporate Pipeline of Oil and Gas Funding into Higher Education." *Humanity and Society, 44*: 151–77.

Lamb, William F., Giulio Mattioli, Sebastian Levi, J. Timmons Roberts, Stuart Capstick, Felix Creutzig, Jan C. Minx, Finn Müller-Hansen, Trevor Culhane, and Julia K. Steinberger. 2020. "Discourses of Climate Delay." *Global Sustainability, 3*: e17.

Lapavitsas, C. 2013. *Profiting without Producing: How Finance Exploits Us All*. Verso.

Legg, Tess, Jenny Hatchard, and Anna B. Gilmore. 2021. "The Science for Profit Model—How and Why Corporations Influence Science and the Use of Science in Policy and Practice." *PLoS One, 16*: e0253272.

Lennon, Erica, and Brian J. Mistler. 2010. "Breaking the Binary: Providing Effective Counseling to Transgender Students in College and University Settings." *Journal of LGBT Issues in Counseling, 4*: 228–40.

Lennon, Myles. 2017. "Decolonizing Energy: Black Lives Matter and Technoscientific Expertise amid Solar Transitions." *Energy Research and Social Science, 30*: 18–27.

Leonard, Christopher. 2019. *Kochland: The Secret History of Koch Industries and Corporate Power in America*. Simon and Schuster.

Leung, Shirley. 2022. "How Harvard Could Spend That $100 Million to Unwind Its Legacies of Slavery." *Boston Globe*.

Levy, David. 2005. "Hegemony in the Global Factory: Power, Ideology, and Value in Global Production Networks." *Academy of Management Proceedings, 2005*: C1–C6.

Lewin, Tamar. 2013. "Pay for U.S. College Presidents Continues to Grow." *New York Times*, December 15.

Lewis, Greg. 2023. "The Powerful Influence of Business on University Boards." Century Foundation. https://tcf.org/content/commentary/the-powerful-influence-of-business-on-university-boards.

Li, Mei, Gregory Trencher, and Jusen Asuka. 2022. "The Clean Energy Claims of BP, Chevron, ExxonMobil and Shell: A Mismatch between Discourse, Actions and Investments." *PLoS One, 17*(2).

Li, Winnie M. 2014. "Dear Harvard: You're Not Winning." *Huffington Post*. https://www.huffpost.com/entry/harvard-rape_b_5079796.

Liboiron, Max. 2023. "What It Means to Practice Values-Based Research." *Nature*. https://www.nature.com/articles/d41586-023-01878-1.

Lieber, Ron. 2021. "Elite Universities Are in an Amenities Arms Race." *Town and Country*. https://www.townandcountrymag.com/society/money-and-power/a35561796/college-luxury-amenities/.

Liu, Catherine. 2023. "Universities Are Turning into Real-Estate Hedge Funds—And Students Are Paying the Price." *Business Insider*. https://www.businessinsider.com/universities-colleges-turning-into-real-estate-hedge-funds-higher-education-2023-3?utm_source=digg&utm_medium=social&utm_campaign=standard-post&r=US&IR=T.

Loorbach, Derk A., and Julia Wittmayer. 2024. "Transforming Universities." *Sustainability Science, 19*: 19–33. https://doi.org/10.1007/s11625-023-01335-y.

Lorek, Sylvia, Kate Power, and Natasha Parker. 2023. *Economies That Dare to Care: Achieving Social Justice and Preventing Ecological Breakdown by Putting Care at the Heart of Our Societies*. Hot or Cool Institute. https://hotorcool.org/wp-content/uploads/2023/07/Economies-that-Dare-to-Care.pdf.

Lueddeke, George Richard. 2020. "Universities in the Early Decades of the Third Millennium: Saving the World from Itself?" In E. Sengupta, P. Blessinger, and C. Mahoney (Eds.), *Civil Society and Social Responsibility in Higher Education: International Perspectives on Curriculum and Teaching Development*. Vol. 21, 229–66. Emerald Publishing.

Lundh, Andreas, Joel Lexchin, Barbara Mintzes, Jeppe B. Schroll and Lisa Bero. 2018. "Industry Sponsorship and research outcome: systematic review with meta-analysis." *Intensive Care Med, 44*(10): 1603–1612.

Machado de Oliveira, Vanessa. 2022. *Hospicing Modernity: Facing Humanity's Wrongs and the Implications for Social Activism*. Penguin Random House.

MacKenzie, Megan, Ozlem Sensoy, Genevieve Fuji Johnson, Nathalie Sinclair, and Laurel Weldon. 2023. "4 Ways Universities Gaslight DEI Initiatives." *Inside Higher Ed*. https://www.insidehighered.com/opinion/views/2023/10/05/how-universities-gaslight-dei-initiatives-opinion.

Mader, P, D Mertens, and N van der Zwan. 2020. *The Routledge International Handbook of Financialization*. Routledge.

Magan, Manchan. 2020. *Thirty-Two Words for Field: Lost Words of the Irish Landscape*. Gill Books.

Malleson, Tom. 2023. *Against Inequality: The Practical and Ethical Case for Abolishing the Superrich*. Oxford University Press.

Manassah, Tala Jamal, Marieke van Woerkom, Jillian Luft, and Ife Lenard. 2022. "Teaching as an Act of Solidarity: A Beginners Guide to Equity in Schools." Morningside Center for Teaching Social Responsibility. https://www.morningsidecenter.org/news/teaching-act-solidarity-introduction.

Manne, Kate. 2018. *Down Girl: The Logic of Misogyny*. Oxford University Press.

Marine, S. B., and Z Nicolazzo. 2014. "Names That Matter: Exploring the Tensions of Campus LGBTQ Centers and Trans* Inclusion." *Journal of Diversity in Higher Education*, 7: 265–81.

Marks, Isaac, and Adolf Tobeña. 1990. "Learning and Unlearning Fear: A Clinical and Evolutionary Perspective." *Neuroscience and Biobehavioral Reviews*, 14: 365–84.

Markusson, Nils, Mads Dahl Gjefsen, Jennie C. Stephens, and David Tyfield. 2017. "The Political Economy of Technical Fixes: The (Mis)alignment of Clean Fossil and Political Regimes." *Energy Research and Social Science*, 23: 1–10.

Matthaei, Julie, and Matthew Slaats. 2023. "The Solidarity Economy: A Way Forward for Our De-futured World." *Journal of Social Encounters*, 7(2): 24–48.

Matthews, Richard. 2021. "Fueling Disinformation: How Big Oil Obstructs Climate Education." *Change Oracle*. https://changeoracle.com/2021/09/20/fueling-disinformation-how-big-oil-obstructs-climate-education/.

Mazzucato, Mariana. 2013. *The Entrepreneurial State*. Anthem Press.

Mazzucato, Mariana. 2020. *Mission Economy: A Moonshot Guide to Changing Capitalism*. Penguin Random House.

Mbembe, A. 2021. *Out of the Dark Night: Essays on Decolonization*. Columbia University Press.

McBride, Ruari-Santiago, and Aoife Neary. 2021. "Trans and Gender Diverse Youth Resisting Cisnormativity in School." *Gender and Education*, 33: 1090–107.

McGeown, Calum, and John Barry. 2023. "Agents of (Un)Sustainability: Democratising Universities for the Planetary Crisis." *Frontiers in Sustainability*, 4.

McKibben, Bill. 2023. "We're in for a Stretch of Heavy Climate." *The Crucial Years* (blog), April 15. https://billmckibben.substack.com/p/were-in-for-a-stretch-of-heavy-climate.

McKie, Ruth E. 2021. "Obstruction, Delay, and Transnationalism: Examining the Online Climate Change Counter-Movement." *Energy Research and Social Science*, 80: 102217.

McLaren, Duncan. 2016. "Mitigation Deterrence and the 'Moral Hazard' of Solar Radiation Management." *Earth's Future*, 4: 596–602.

McMahon, Cian. 2020. "Review Essay: The Case for Community Wealth Building." *Community Development Journal, 56*: 356–62.

Mehra, Bharat, and Rebecca Davis. 2015. "A Strategic Diversity Manifesto for Public Libraries in the 21st Century." *New Library World, 116*(1/2): 15–36. https://doi.org/10.1108/NLW-04-2014-0043.

Meiksins Wood, Ellen. 2017. *The Origin of Capitalism: A Longer View*. Verso Books.

Michael, Nancy, and Ben Wilson. 2021. "Unlearning Is the New Learning: A Neuroscientific and Theological Case for How and Why to See the World Differently." *Church Life Journal*. https://churchlifejournal.nd.edu/articles/unlearning-is-the-new-learning-a-neuroscientific-and-theological-case-for-how-and-why-to-see-the-world-differently/.

Michaels, David. 2020. *The Triumph of Doubt: Dark Money and the Science of Deception*. Oxford University Press.

Mignolo, Walter D. 2009. "Epistemic Disobedience, Independent Thought and Decolonial Freedom." *Theory, Culture and Society, 26*: 159–81.

Mikkelson, Gregory M., Miron Avidan, Aleksandra Conevska, and Dror Etzion. 2021. "Mutual Reinforcement of Academic Reputation and Fossil Fuel Divestment." *Global Sustainability, 4*: e20.

Molina, Olga, and George A. Jacinto. 2015. "The Advantages and Benefits of a Student Mutual-Aid Group in Developing Groupwork Skills." *Groupwork, 25*: 78–92.

Mondragon Unibertsitatea. 2023. "Cooperative University." Accessed June 22, 2023.

Moody, Josh. 2020. "10 Colleges Where the Most Alumni Donate." *U.S. News and World Report*, December 8.

Morris, Viveca, and Jennifer Jacquet. 2024. "The Animal Agriculture Industry, U.S. University Centers and Professors, and the Obstruction of Climate Understanding and Policy." *Climatic Change, 177*(3): 41.

Moses, Michele S., and Laura Dudley Jenkins. 2014. "Affirmative Action Should Be Viewed in Global Context." *Conversation*. https://theconversation.com/affirmative-action-should-be-viewed-in-global-context-33618.

Mtawa, Ntimi N., and Gerald Wangenge-Ouma. 2022. "Questioning Private Good Driven University-Community Engagement: A Tanzanian Case Study." *Higher Education, 83*: 597–611.

NASEM. 2021. "Reflecting Sunlight: Recommendations for Solar Geoengineering Research and Research Governance." Consensus Study Report by the National Academies of Sciences, Engineering, and Medicine. https://nap.nationalacademies.org/catalog/25762/reflecting-sunlight-recommendations-for-solar-geoengineering-research-and-research-governance.

National Academies of Sciences, Engineering, and Medicine. 2018. *Sexual Harassment of Women: Climate, Culture, and Consequences in Academic Sciences, Engineering, and Medicine*. National Academies Press.

Nayak, Preeti. 2023. "Thinking with Climate Coloniality in the Classroom: Possibilities for Climate Justice Education." American Association of Geography Annual Conference.

Nelson, Julie A. 1996. *Feminism, Objectivity and Economics*. Routledge.

Nemes, Noémi, Stephen J. Scanlan, Pete Smith, Tone Smith, Melissa Aronczyk, Stephanie Hill, Simon L. Lewis, A. Wren Montgomery, Francesco N. Tubiella, and Doreen Stabinsky. 2022. "An Integrated Framework to Assess Greenwashing." *Sustainability*, 14(8): 4431. https://www.mdpi.com/2071-1050/14/8/4431.

Newell, Peter, and Matthew Paterson. 2010. *Climate Capitalism: Global Warming and the Transformation of the Global Economy*. Cambridge University Press.

Newell, Peter, and Andrew Simms. 2020. "Towards a Fossil Fuel Non-Proliferation Treaty." *Climate Policy*, 20: 1043–54.

Newell, Peter, Shilpi Srivastava, Lars Otto Naess, Gerardo A. Torres Contreras, and Roz Price. 2021. "Toward Transformative Climate Justice: An Emerging Research Agenda." *WIREs Climate Change*, 12: e733.

Newell, Peter, Harro van Asselt, and Freddie Daley. 2022. "Building a Fossil Fuel Non-Proliferation Treaty: Key Elements." *Earth System Governance*, 14: 100159.

Newman, John Henry. 1893. *The Idea of a University*. Longmans Green.

Nishitani, Kimitaka, Thi Bich Hue Nguyen, Trong Quy Trinh, Qi Wu, and Katsuhiko Kokubu. 2021. "Are Corporate Environmental Activities to Meet Sustainable Development Goals (SDGs) Simply Greenwashing? An Empirical Study of Environmental Management Control Systems in Vietnamese Companies from the Stakeholder Management Perspective." *Journal of Environmental Management*, 296: 113364.

Northeastern. 2023. "Voices of Northeastern Alumni: Edward G. Galante E'73." Accessed June 19, 2023. https://voices.northeastern.edu/galante/index.html.

Nunes, Ludmila. 2021. "New Directions for Diversity, Equity and Inclusion in Higher Education." Association for Psychological Science. https://www.psychologicalscience.org/observer/words-to-action.

O'Hara, Glen. 2023. "How Can England's Universities Survive?" *Guardian*, March 27. https://www.theguardian.com/books/2023/mar/27/the-big-idea-how-can-englands-universities-survive.

Oksala, Johanna. 2023. *Feminism, Capitalism, and Ecology*. Northwestern University Press.

Oreskes, Naomi. 2015. "Universities Must Divest from the Fossil Fuel Industry." *New York Times*, August 8.

Oreskes, Naomi, and Eric M. Conway. 2010. *Merchants of Doubt: How a Handful of Scientists Obscured the Truth on Issues from Tobacco Smoke to Climate Change*. Bloomsbury.

Oreskes, Naomi, and Erik M. Conway. 2023. *The Big Myth: How American Business Taught Us to Loathe Government and Love the Free Market*. Bloomsbury.

Orlove, Ben, Pasang Sherpa, Neil Dawson, Ibidun Adelekan, Wilfredo Alangui, Rosario Carmona, Deborah Coen, Melissa K. Nelson, Victoria Reyes-García, Jennifer Rubis, Gideon Sanago, and Andrew Wilson. 2023. "Placing Diverse Knowledge Systems at the Core of Transformative Climate Research." *Ambio*.

Overland, Indra, and Benjamin K Sovacool. 2020. "The Misallocation of Climate Research Funding." *Energy Research and Social Science, 62*: 101349.

Owen, Bruce M, and Ronald Braeutigam. 1978. *Regulation Game: Strategic Use of the Administrative Process*. Harper Collins.

Papadimitriou, Antigoni, and Marius Boboc. (eds.). 2021. *Re-envisioning Higher Education's Public Mission*: Palgrave Macmillan.

paperson, la. 2017. *A Third University Is Possible*. University of Minnesota Press.

Paraskeva, João M. 2017. "Against the Epistemicide. Itinerant Curriculum Theory and the Reiteration of an Epistemology of Liberation." In Michael Uljens and Rose M. Ylimaki (Eds.), *Bridging Educational Leadership, Curriculum Theory and Didaktik: Non-affirmative Theory of Education*. Springer International Publishing.

Parkes, Susan M. 2004. *A Danger to the Men? A History of Women in Trinity College Dublin 1904–2004*. Lilliput Press.

Patel, Leigh. 2021. *No Study without Struggle: Confronting Settler Colonialism in Higher Education*. Beacon Press.

Peer, Verena, and Gernot Stoeglehner. 2013. "Universities as Change Agents for Sustainability—Framing the Role of Knowledge Transfer and Generation in Regional Development Processes." *Journal of Cleaner Production, 44*: 85–95.

Pel, Bonno, Julia M. Wittmaver, Flor Avelino, and Tom Bauler. 2023. "Paradoxes of Transformative Social Innovation: From Critical Awareness towards Strategies of Inquiry." *Novation: Critical Studies of Innovation*. https://revistas.ufpr.br/novation/article/view/91113.

Perez, Carlota. 2008. "ICT and Green. A Natural Partnership to Nurture. Shaping a Sustainable Globalization and Redesigning the 'Good Life.'" In *Connected Urban Development Global Conference*. Amsterdam. http://

www.carlotaperez.org/lecturesandvideos.html_https://exchange.clarku.edu/owa/redir.aspx?C=b016aa5c16344b609350ebeeef8363c0&URL=http%3a%2f%2fwww.carlotaperez.org%2flecturesandvideos.html.

Perez-Lugo, Marla, Cecilio Ortiz Garcia, and D. Valdes. 2021. "Understanding Hurricane Maria through Puerto Rico's Electrical System: Disaster Response as Transition Management." In M. T. Mora, H. Rodriguez, and A. Davila (Eds.), *Hurricane Maria in Puerto Rico; Disaster, Vulnerability and Resilience*. Lexington Books.

Perry, Mia. 2024. *Pluriversal Literacies for Sustainable Futures: When Words Are Not Enough*. Routledge.

Piketty, Thomas. 2015. *The Economics of Inequality*. Harvard University Press.

Pinto, Manuela F. 2015. "Tensions in Agnotology: Normativity in the Studies of Commercially Driven Ignorance." *Social Studies of Science*, 45(2): 294–315. http://www.jstor.org/stable/43829025.

Pirgmaier, Elke, and Julia K. Steinberger. 2019. "Roots, Riots, and Radical Change—A Road Less Travelled for Ecological Economics." *Sustainability*, 11: 2001.

Porter, Sha-shonda, John Wang, and Shannon Dunn. 2023. "Tackling Systemic and Structural Inequities in Higher Ed IT: A Primer on Beginnings." *Educause Review*. https://er.educause.edu/articles/2023/4/tackling-systemic-and-structural-inequities-in-higher-ed-it-a-primer-on-beginnings.

Positive Money. 2018. "A Green Bank of England: Central Banking for a Low-Carbon Economy." Positive Money. http://positivemoney.org/wp-content/uploads/2018/05/PositiveMoney_AGreenBankofEngland_Web.pdf.

Positive Money Europe. 2020. "The ECB and Climate Change: Outlining a Vision for Success." *Policy Briefing*. Positive Money Europe with New Economics Foundation and 350.org. https://www.positivemoney.eu/publications/.

Powell, Jerome. 2023. "Conversation with the Chair: A Techer Town Hall Meeting." Federal Reserve. https://www.federalreserve.gov/conferences/chair-powell-teacher-town-hall-2023.htm.

Pritchard, Erin, and Delyth Edwards (Eds.). 2023. *Sexual Misconduct in Academia: Informing an Ethics of Care in the University*. Routledge.

Pritchett, Lant, and Addison Lewis. 2022. "Economic Growth Is Enough and Only Economic Growth Is Enough." https://lantpritchett.org/wp-content/uploads/2022/05/Basics-legatum-paper_short.pdf.

Proctor, Robert N., and Londa Schiebinger. (Eds.). 2008. *Agnotology: The Making and Unmaking of Ignorance*. Stanford University Press.

Quadlin, Natasha, and Brian Powell. 2022. *Who Should Pay? Higher Education, Responsibilty, and the Public*. Russell Sage.

Quan, Melissa. 2023. "Framework for Justice-Centering Relationships: Implications for Place-Based Pedagogical Practice." *Metropolitan Universities*, 34(2): 138–157. https://doi.org/10.18060/26450.

Quigley, Julia. 2018. "Payments in Lieu of Trouble: Nonprofit Pilots as Extortion or Efficient Public Finance?" *NYU Environmental Law Journal, 26*.

Rabe, Chris, et al. 2023. *Climate Justice Instructional Toolkit*. MIT Environmental Solutions Initiative. https://environmentalsolutions.mit.edu/climate-justice-instructional-toolkit/.

Ramanujam, Archana. 2023. "Climate Scholarship Needs Du Bois: Climate Crisis through the Lens of Racial and Colonial Capitalism*," *Sociological Inquiry, 93*: 273–95.

Raworth, Kate. 2017. *Doughnut Economics: Seven Ways to Think like a 21st-Century Economist*. Random House.

Renner, K. Edward, and Thom Moore. 2004. "The More Things Change, the More They Stay the Same: The Elusive Search for Racial Equity in Higher Education." *Analyses of Social Issues and Public Policy (ASAP), 4*: 227–41.

Renouard, Cécile, Rémi Beau, Christophe Goupil, and Christian Koenig (Eds.). 2021. *The Great Transition Guide: Principles for a Transformative Education*. Oxford University Press.

Riccio, Rebecca, Giordana Mecagni, and Becca Berkey. 2022. "Principles of Anti-Oppressive Community Engagement for University Educators and Researchers." Social Impact Lab, Northeastern University.

Riley, Naomi Schaefer, and James Piereson. 2020. "Reimaging the Public University." *National Affairs*. https://www.nationalaffairs.com/publications/detail/reimagining-the-public-university.

RISE. 2023. "Welcome to the RISE Network." Accessed June 12, 2023. https://therisenetwork.org/.

Roberts-Gregory, Frances. 2021. "Climate Justice in the Wild n' Dirty South: An Autoethnographic Reflection on Ecowomanism as Engaged Scholar-Activist Praxis before and during COVID-19." In K. Melchor Quick Hall and Gwyn Kirk (Eds.), *Mapping Gendered Ecologies: Engaging with and Beyond Ecowomanism and Ecofeminism*. Lexington Books.

Robinson, Jenna A. 2017. "The Bennett Hypothesis Turns 30." The James G. Martin Center for Academic Renewal. https://files.eric.ed.gov/fulltext/ED588382.pdf.

Robra, Ben, Alex Pazaitis, Chris Giotitsas, and Mario Pansera. 2023. "From Creative Destruction to Convivial Innovation—A Post-Growth Perspective." *Technovation, 125*: 102760.

Rockström, Johan, Joyeeta Gupta, Dahe Qin, Steven J. Lade, Jesse F. Abrams, Lauren S. Andersen, David I. Armstrong McKay, et al. 2023. "Safe and Just Earth System Boundaries." *Nature*.

Røpke, Inge. 2020. "Econ 101—In Need of a Sustainability Transition." *Ecological Economics*, 169: 106515.

Roth, Laura, Bertie Russell, and Matthew Thompson. 2023. "Politicising Proximity: Radical Municipalism as a Strategy in Crisis." *Urban Studies*, 60(11): 2009–35. https://doi.org/10.1177/00420980231173825.

Rowlands, Julie. 2015. "Turning Collegial Governance on Its Head: Symbolic Violence, Hegemony and the Academic Board." *British Journal of Sociology of Education*, 36: 1017–35.

Royal Irish Academy. 2021. *The Role of Regions and Place in Higher Education across the Island of Ireland*. Royal Irish Academy. Higher Education Futures Taskforce.

Rudy, Willis. 1984. *The Universities of Europe, 1100–1914: A History*. Associated University Presses.

Russel, Dominic, Alan Smith, and Carrie Sloan. 2016. "The Financialization of Higher Education." Roosevelt Institute.

Sareen, Siddharth, and Katinka Lund Waagsaether. 2022. "New Municipalism and the Governance of Urban Transitions to Sustainability." *Urban Studies*, 60(11): 2271–89. https://doi.org/10.1177/00420980221114968.

Scholars at Risk Network. 2023. "Peace Petition Scholars, Turkey." https://www.scholarsatrisk.org/actions/academics-for-peace-turkey/.

Schoorman, Dilys. 2018. "The Erosion of Faculty Governance." *Counterpoints*, 517: 237–51.

SDG Academy. 2023. "Free Educational Resources from the World's Leading Experts on Sustainable Development." https://sdgacademy.org/.

Sederstrom, Nneka, and Tamika Lasege. 2022. "Anti-Black Racism as a Chronic Condition." In Faith E. Fletcher and et al. (Eds.), *A Critical Moment in Bioethics: Reckoning with Anti-Black Racism through Intergenerational Dialogue*. Hastings Center Report 52. https://onlinelibrary.wiley.com/doi/pdf/10.1002/hast.1364.

Servant-Miklos, Virginie. 2024. *Pedagogies of Collapse: A Hopeful Education for the End of the World as We Know It*. Bloomsbury Academic.

Shahjahan, Riyad A., Annabelle L. Estera, Kristen L. Surla, and Kirsten T. Edwards. 2022. "'Decolonizing' Curriculum and Pedagogy: A Comparative Review across Disciplines and Global Higher Education Contexts." *Review of Educational Research*, 92: 73–113.

Shaibah, Arig al 2023. *Advancing Educational and Social Equity in and Through Higher Education: Lessons from a Senior Equity Administrator in the Canadian Context*. Paper presented at the 17th International Conference for Higher Educational Reform, Glasgow, Scotland, June 19–23.

Si, Yutong, Dipa Desai, Diana Bozhilova, Sheila Puffer, and Jennie C. Stephens. 2023. "Fossil Fuel Companies' Climate Communication Strategies: Industry Messaging on Renewables and Natural Gas." *Energy Research and Social Science, 98*.

Silva, Elsa Costa e. 2014. "Crisis, Financialization and Regulation: The Case of Media Industries in Portugal." *The Political Economy of Communication*, 2(2): https://www.polecom.org/index.php/polecom/article/view/38/236.

Simmons, Holiday, and Fresh White. 2014. "Our Many Selves." In L Erickson-Schroth (Ed.), *Trans Bodies, Trans Selves: A Resource for the Transgender Community*. Oxford University Press.

Singer, Merrill. 2018. *Climate Change and Social Inequality: The Health and Social Costs of Global Warming*. Routledge.

Slaughter, Sheila, and Larry L. Leslie. 1999. *Academic Capitalism: Politics, Policies, and the Entrepreneurial University*. Johns Hopkins University Press.

Smith, David M. 1998. "How Far Should We Care? On the Spatial Scope of Beneficence." *Progress in Human Geography*, 22(1), 15–38. https://doi.org/10.1191/030913298670636601.

Smith, David M. 1999. "Geography and Ethics: How Far Should We Go?" *Progress in Human Geography*, 23: 119–25.

Sokol, Martin. 2003. "Regional Dimensions of the Knowledge Economy: Implications for the 'New Europe.'" PhD thesis, University of Newcastle upon Tyne, England.

Sokol, Martin. 2011. *Economic Geographies of Globalisation*: Edward Elgar.

Sokol, Martin, and Jennie C. Stephens. 2022. "Monetary Policy and Ecological Crisis: Towards a Climate Justice Approach." In *Paper Presented at 26th Forum for Macroeconomics and Macroeconomic Policy (FMM) Conference on "Post-Keynesian Economics and Global Challenges."* Berlin, Germany, October 20–22.

Solnit, Rebecca, and Thelma Young Lutunatabua (Eds.). 2023. *Not Too Late: Changing the Climate Story from Despair to Possibility*. Haymarket Books.

Song, Lily, and Elifmina Mizrahi. 2023. "From Infrastructural Repair to Reparative Planning." *Journal of the American Planning Association*: 1–14.

Speth, James Gustave. 2008. *The Bridge at the Edge of the World: Capitalism, the Environment, and Crossing from Crisis to Sustainability*. Yale University Press.

Starr, G. Gabrielle. 2023. "AI Can Enhance the Pleasures of Learning." *Chronicle of Higher Education*. https://www.chronicle.com/article/how-will-artificial-intelligence-change-higher-ed?utm_source=Iterable&utm_medium=email&utm_campaign=campaign_6923362_nl_Academe-Today_date_20230530&cid=at&source=ams&sourceid=.

St. Clair, Ralf. 2020. *Learning-Centered Leadership in Higher Education*. Palgrave Macmillan.

Steele, Wendy, and Lauren Rickards. 2021. *The Sustainable Development Goals in Higher Education: A Transformative Agenda?* Springer.

Steffen, Will, Katherine Richardson, Johan Rockström, Sarah E. Cornell, Ingo Fetzer, Elena M. Bennett, Reinette Biggs, et al. 2015. "Planetary Boundaries: Guiding Human Development on a Changing Planet." *Science*, 347: 1259855.

Stein, Sharon. 2022. *Unsettling the University: Confronting the Colonial Foundations of US Higher Education*. Johns Hopkins University Press.

Stephens, Jennie C. 2009. "Technology Leader, Policy Laggard: Carbon Capture and Storage (CCS) Development for Climate Mitigation in the U.S. Political Context." In James Meadowcroft and Oluf Langhelle (Eds.), *Caching the Carbon: The Politics and Policy of Carbon Capture and Storage*. Edward Elgar.

Stephens, Jennie C. 2014. "Time to Stop Investing in Carbon Capture and Storage and Reduce Government Subsidies of Fossil-Fuels." *Wiley Interdisciplinary Reviews: Climate Change*, 5: 169–73.

Stephens, Jennie C. 2020. *Diversifying Power: Why We Need Antiracist, Feminist Leadership on Climate and Energy*. Island Press.

Stephens, Jennie C. 2022. "Beyond Climate Isolationism: A Necessary Shift for Climate Justice." *Current Climate Change Reports*, 8: 83–90.

Stephens, Jennie C. 2023. "Reconnecting Economics Education with Today's Global Realities." *Nonprofit Quarterly*. https://nonprofitquarterly.org/reconnecting-economics-education-with-todays-global-realities/.

Stephens, Jennie C. 2024. "Divesting from Israel's War Isn't Naive. Students Did It with Fossil Fuels." *Truthout*. https://truthout.org/articles/divesting-from-israels-war-isnt-naive-students-did-it-with-fossil-fuels/.

Stephens, Jennie C., Peter Frumhoff, and Leehi Yona. 2018. "The Role of College and University Faculty in the Fossil Fuel Divestment Movement." *Elementa: Science of the Anthropocene*, 6(1): 41. https://doi.org/10.1525/elementa.297.

Stephens, Jennie C., and Amanda C. Graham. 2010. "Toward an Empirical Research Agenda for Sustainability in Higher Education: Exploring the Transition Management Framework." *Journal of Cleaner Production*, 18: 611–18.

Stephens, Jennie C., Prakash Kashwan, Duncan McLaren, and Kevin Surprise. 2021. "The Dangers of Mainstreaming Solar Geoengineering: A Critique of the National Academies Report." *Environmental Politics*: 1–10.

Stephens, Jennie C., and Nils Markusson. 2018. "Technological Optimism in Climate Mitigation: The Case of Carbon Capture and Storage." In Matthias Gross and Debra J. Davidson (Eds.), *Oxford Handbook of Energy and Society*. Oxford University Press.

Stephens, Jennie C., and Martin Sokol. 2023. "Financial Innovation for Climate Justice: Central Banks and Transformative 'Creative Disruption.'" *Climate and Development*: 1–12.

Stephens, Jennie C., and Kevin Surprise. 2020. "The Hidden Injustices of Advancing Solar Geoengineering Research." *Global Sustainability, 3*: 1–6.

Stephens, Jennie C., E. J. Wilson, and T. R. Peterson. 2015. *Smart Grid (R) Evolution: Electric Power Struggles*. Cambridge University Press.

Sterling, S., and I. Thomas. 2006. "Education for Sustainability: The Role of Capabilities in Guiding University Curricula." *International Journal of Innovation and Sustainable Development, 1*: 349–70.

Stewart, Abigail J., and Virginia Valian. 2018. *An Inclusive Academy: Achieving Diversity and Excellence*. MIT Press.

Stewart, James B. 2023. "Tax Me If You Can." *New Yorker*.

Sullivan, Margaret. 2020. *Ghosting the News: Local Journalism and the Crisis of American Democracy*. Columbia Global Reports.

Sultana, Farhana. 2022a. "Critical Climate Justice." *Geographical Journal, 188*: 118–24.

Sultana, Farhana. 2022b. "The Unbearable Heaviness of Climate Coloniality." *Political Geography, 99*: 102638.

Sultana, Farhana (Ed). 2024. *Confronting Climate Coloniality: Decolonizing Pathways for Climate Justice*. Routledge.

Supran, Geoffrey, Stefan Rahmstorf, and Naomi Oreskes. 2023. "Assessing ExxonMobil's Global Warming Projections." *Science, 379*.

Suresh Babu, GS. 2023. "Adaptation in the Scientific Method: An Outline for Mutual Learning and Knowledge Co-production in Climate Science." *Science, Technology and Society, 28*: 621–38.

Surprise, Kevin. 2020. "Stratospheric Imperialism: Liberalism, (Eco)Modernization, and Ideologies of Solar Geoengineering Research." *Environment and Planning E: Nature and Space, 3*: 141–63.

Tabuchi, Hiroko. 2022. "Kicking Oil Companies Out of School." *New York Times*, August 16.

Táíwò, Olúfẹ́mi O. 2023. "A Framework to Help Us Better Understand the World." *Hammer and Hope: A Magazine of Black Politics and Culture, 1*. https://hammerandhope.org/article/issue-1-article-8.

Táíwò, Olúfẹ́mi O., and Beba Cibralic. 2021. "The Case for Climate Reparations." *Foreign Policy*.

Tamale, Sylvia. 2022. *Decolonization and Afro-Feminism*. Daraja Press.

Tang, Li, and Hugo Horta. 2021. "Women Academics in Chinese Universities: A Historical Perspective." *Higher Education, 82*: 865–95.

Tasci, Ufuk Necat. 2020. "The World's First, and Still-Operating, University Is in Morocco." *TRT World*. https://www.trtworld.com/magazine/the-world-s-first-and-still-operating-university-is-in-morocco-36652.

Taylor, Dorceta E. 2018. "Diversity in Environmental Organizations: Reporting and Transparency." University of Michigan, School for Environment and Sustainability.

Teichert, Sebastian, Martin G. J. Löder, Ines Pyko, Marlene Mordek, Christian Schulbert, Max Wisshak, and Christian Laforsch. 2021. "Microplastic Contamination of the Drilling Bivalve Hiatella arctica in Arctic Rhodolith Beds." *Scientific Reports*, 11: 14574.

Tejani, Sheba. 2019. "What's Feminist about Feminist Economics?" *Journal of Economic Methodology*, 26: 99–117.

Tenbrunsel, Ann E., McKenzie R. Rees, and Kristina A. Diekmann. 2019. "Sexual Harassment in Academia: Ethical Climates and Bounded Ethicality." *Annual Review of Psychology*, 70: 245–70.

Thaller, Annina, Anna Schreuer, and Alfred Posch. 2021. "Flying High in Academia—Willingness of University Staff to Perform Low-Carbon Behavior Change in Business Travel." *Frontiers in Sustainability*, 2. https://doi.org/10.3389/frsus.2021.790807.

The Scope. 2023. "About." https://thescopeboston.org/about-2/.

Tierney, Kathleen, Christine Bevc, and Erica Kuligowski. 2006. "Metaphors Matter: Disaster Myths, Media Frames, and Their Consequences in Hurricane Katrina." *Annals of the American Academy of Political and Social Science*, 604: 57–81.

Tigue, Kristoffer. 2023. "Climate Change Made the Texas Heat Wave More Intense. Renewables Softened the Blow." *Inside Climate News*. https://insideclimatenews.org/news/23062023/todays-climate-texas-heat-climate-renewables/.

Tillman, Linda C. 2018. "Achieving Racial Equity in Higher Education: The Case for Mentoring Faculty of Color." *Teachers College Record*, 120: 1–18.

Toffler, Alvin. 1990. *Powershift: Knowledge, Wealth, and Power at the Edge of the 21st Century*. Bantam Books.

Tooze, Adam. 2022. "Welcome to the World of the Polycrisis: Today Disparate Shocks Interact So That the Whole Is Worse than the Sum of the Parts." *Financial Times*, October 28. https://www.ft.com/content/498398e7-11b1-494b-9cd3-6d669dc3de33.

Towl, Graham, and Tammi Walker. 2019. *Tackling Sexual Violence at Universities: An International Perspective*. Routledge.

Trinity College Dublin. 2023. "College Governance." Accessed June 26, 2023. https://www.tcd.ie/Secretary/governance/.

Trisos, Christopher H., Giuseppe Amatulli, Jessica Gurevitch, Alan Robock, Lili Xia, and Brian Zambri. 2018. "Potentially Dangerous Consequences for Biodiversity of Solar Geoengineering Implementation and Termination." *Nature Ecology and Evolution*, 2: 475–82.

Tseng, Sherry H. Y., Craig Lee, and James Higham. 2022. "Managing Academic Air Travel Emissions: Towards System-Wide Practice Change." *Transportation Research Part D: Transport and Environment*, 113: 103504. https://doi.org/10.1016/j.trd.2022.103504.

Tuck, Eve. 2009. "Suspending Damage: A Letter to Communities." *Harvard Educational Review*, 79.

Tuck, Eve, and K. Wayne Yang. 2012. "Decolonization Is Not a Metaphor." *Decolonization: Indigeneity, Education and Society*, 1. https://jps.library.utoronto.ca/index.php/des/article/view/18630.

Tuck, Eve, and K. Wayne Yang. 2014. "Unbecoming Claims: Pedagogies of Refusal in Qualitative Research." *Qualitative Inquiry*, 20: 811–18.

Turnheim, Bruno, and Frank W. Geels. 2013. "The Destabilization of Existing Regimes: Confronting a Multi-dimensional Framework with a Case Study of the British Coal Industry (1913–1967)." *Research Policy*, 42: 1749–67.

UCS. 2023. "The Fossil Fuels behind Forest Fires: Quantifying the Contribution of Major Carbon Producers to Increasing Wildfire Risk in Western North America." Union of Concerned Scientists. https://www.ucsusa.org/sites/default/files/2023-05/fossil-fuels-behind-forest-fires-full-report-2023_0.pdf.

UKRI. 2023. "SUS-POL: Supply-Side Policies for Fossil Fuels." United Kingdom Research Institute. Accessed June 20, 2023. https://gtr.ukri.org/projects?ref=EP%2FX035964%2F1.

UNESCO. 2020. *Towards Universal Access to Higher Education: International Trends*. https://unesdoc.unesco.org/ark:/48223/pf0000375686.locale=en.

UNESCO. 2022. "Knowledge-Driven Actions: Transforming Higher Education for Global Sustainability." UNESCO Digital Library. https://unesdoc.unesco.org/ark:/48223/pf0000380519.

United Frontline Table. 2022. *A People's Orientation to a Regenerative Economy: Protect, Repair, Invest, and Transform*. https://unitedfrontlinetable.org/report/.

United Nations. 2019. "Unprecedented Impacts of Climate Change Disproportionately Burdening Developing Countries, Delegate Stresses, as Second Committee Concludes General Debate." United Nations Meetings Coverages and Press Releases.

UnKoch My Campus. 2020. "UnKoch My Campus." https://www.unkochmycampus.org/.

Unlearning Economics. 2023. "Unlearning Economics." YouTube channel. https://www.youtube.com/@unlearningeconomics9021.

Urai, Anne E., and Clare Kelly. 2023. "Rethinking Academia in a Time of Climate Crisis." *eLife, 12*: e84991.

Vakulchuk, Roman, and Indra Overland. 2024. "The Failure to Decarbonize the Global Energy Education System: Carbon Lock-In and Stranded Skill Sets." *Energy Research and Social Science 110*: 103446.

Valantine, Hannah A., and Francis S. Collins. 2015. "National Institutes of Health Addresses the Science of Diversity." *Proceedings of the National Academy of Sciences, 112*: 12240–42.

Valimaa, Jussi. 2015. "Why Finland and Norway Still Shun University Tuition Fees—Even for International Students." *Conversation*.

van Damme, Dirk. 2021. "Transforming Universities for a Sustainable Future." In Hilligje van't Land, Andreas Corcoran, and Diana-Camelia Iancu (Eds.), *The Promise of Higher Education: Essays in Honour of 70 Years of IAU*. Springer International Publishing.

van den Berg, Bas. 2023. "Design Principles for Regenerative Higher Education in Times of Sustainability Transitions." PhD thesis, Wageningen University.

van Oers, Laura, Giuseppe Feola, Hens Runhaar, and Ellen Moors. 2023. "Unlearning in Sustainability Transitions: Insight from Two Dutch Community-Supported Agriculture Farms." *Environmental Innovation and Societal Transitions, 46*: 100693.

Viaene, Liesolotte, Catarina Laranjeiro, and Miye Nadya Tom. 2023. "The Walls Spoke When No One Else Would: Autoethnographic Notes on Sexual-Power Gatekeeping within Avant-Garde Academia." In Erin Pritchard and Delyth Edwards (Eds.), *Sexual Misconduct in Academia: Informing an Ethics of Care in the University*. Routledge.

Waddock, Sandra. 2023. *Catalyzing Transformation: Making System Change Happen*. Business Expert Press.

Wagner, Anne, Sandra Acker, and Kimine Mayuzumi. 2008. *Whose University Is It, Anyway? Power and Privilege in Gendered Terrain*. Canadian Scholars.

Wagner, Anne, and June Ying Yee. 2011. "Anti-Oppression in Higher Education: Implicating Neo-Liberalism." *Canadian Social Work Review/Revue canadienne de service social, 28*: 89–105.

Waldron, Fionnuala, Benjamin Mallon, Maria Barry, and Gabriela Martinez Sainz. 2020. "Climate Change Education in Ireland: Emerging Practice in a Context of Resistance." In David Robbins, Diarmuid Torney and Pat Brereton (Eds.), *Ireland and the Climate Crisis*. Springer International Publishing.

Washburn, Jennifer. 2005. *University Inc. The Corporate Corruption of Higher Education*. Basic Books.

Watermeyer, Richard, and Gemma Derrick. 2022. "Why the Party Is Over for

Britain's Research Excellence Framework." *Nature*. https://www.nature.com/articles/d41586-022-01881-y.
Watts, Nick, Markus Amann, Nigel Arnell, Sonja Ayeb-Karlsson, Jessica Beagley, Kristine Belesova, Maxwell Boykoff, Peter Byass, Wenjia Cai, and Diarmid Campbell-Lendrum. 2021. "The 2020 Report of the Lancet Countdown on Health and Climate Change: Responding to Converging Crises." *Lancet, 397*: 129–70.
Weinberg, Alvin M. 1967. "Can Technology Replace Social Engineering?" *American Behavioral Scientist*: 7–10.
Weinrub, A, and A Giancatarino. 2015. "Toward a Climate Justice Energy Platform: Democratizing Our Energy Future." Local Clean Energy Alliance/Center for Social Inclusion. https://localcleanenergy.org/files/Climate%20Justice%20Energy%20Platform.pdf.
Weiss, Marie, Matthias Barth, Arnim Wiek, and Henrik von Wehrden. 2021. "Drivers and Barriers of Implementing Sustainability Curricula in Higher Education—Assumptions and Evidence." *Higher Education Studies, 11*: 42.
Wenderlich, Michelle Cole. 2021. "Climate Municipalism: Attempts for Politics and Commons through Energy Municipalization Campaigns in Berlin and Minneapolis." PhD thesis. Clark University. https://www.proquest.com/openview/c5e889953285096223c96702d1417c4c/1?pq-origsite=gscholar&cbl=18750&diss=y.
Westervelt, Amy. 2021. "If You Fund the Research, You Can Shape the World." *The Nation*. https://www.thenation.com/article/environment/university-oil-influence/.
Westervelt, Amy. 2023. "Fossil Fuel Companies Donated $700M to US Universities over 10 Years." *Guardian*, March 1.
White House. 2022. "A Proclamation on National Historically Black Colleges and Universities Week." September 16. https://www.whitehouse.gov/briefing-room/presidential-actions/2022/09/16/a-proclamation-on-national-historically-black-colleges-and-universities-week-2022/.
White-Newsome, Jalonne Lynay. 2021. "Experts: Why Does Climate Justice Matter." *Carbon Brief: Clear on Climate*. https://www.carbonbrief.org/experts-why-does-climate-justice-matter/.
Wiksell, Kristin. 2020. "Worker Cooperatives for Social Change: Knowledge-Making through Constructive Resistance within the Capitalist Market Economy." *Journal of Political Power, 13*: 201–16.
Williams, J. J. 2021. "Who's Responsible for Student Debt? The One Percent Deserve Much of the Blame." *Salon*, August 14. https://www.salon.com/2021/2008/2014/whos-responsible-for-student-debt-the-one-percent-deserve-much-of-the-blame/.

Wills, Jane. 2016. *Locating Localism: Statecraft, Citizenship and Democracy*. Bristol University Press.

Wilson, Ralph, and Isaac Kamola. 2021. *Free Speech and Koch Money: Manufacturing a Campus Culture War*. Pluto Press.

Wittmayer, Julia, and Derk Loorbach. In progress. "DRIFT: A Case Study."

Worthen, Molly. 2023. "Why Universities Should Be More like Monasteries." *New York Times*, May 25.

Yacovone, Donals. 2022. *Teaching White Supremacy: America's Democratic Ordeal and the Forging of Our National Identity*. Pantheon Books.

Yeh, James. 2020. "Interview with Robin Wall Kimmerer: People Can't Understand the World as a Gift Unless Someone Shows Them How." *Guardian*, May 23. https://www.theguardian.com/books/2020/may/23/robin-wall-kimmerer-people-cant-understand-the-world-as-a-gift-unless-someone-shows-them-how.

Young, Elise. 2016. "Princeton Will Pay $18 Million to Settle Residents' Tax Case." *Bloomberg*. https://www.bloomberg.com/news/articles/2016-10-15/princeton-will-pay-18-million-to-settle-residents-tax-case?leadSource=uverify%20wall.

Zelikova, Jane, Kelly Ramirez, and Jewel Lipps. 2018. "Harassment Charges: Enough Himpathy." *Science, 361*: 655–55.

Zhao, Jiayi, and Karen Jones. 2017. "Women and Leadership in Higher Education in China: Discourse and the Discursive Construction of Identity." *Administrative Sciences, 7*: 21.

Zinn, Howard. 2002. *You Can't be Neutral on a Moving Train*. Beacon Press.

INDEX

Figures and tables are indicated by "f" and "t" following page numbers.

academic capitalism, 20, 80, 159–68, 178, 181
academic freedom, 22, 46, 56, 61, 80, 222, 234–35
academic-industrial complex, 81–82
accountability: in ADAPT-ing framework, 31, 236; for climate damages, 140; climate justice and, 35; community-engaged research and, 144; culture of, 20, 167, 199, 229–30, 237, 238; importance of, 91; for oppression, 149; power and, 56; public good and, 52; self-reflection and, 12, 72; for sexual harassment, 59; social footprint mapping and, 20, 199
ADAPT-ing framework, 31, 236
African Americans. *See* Black Americans
Agassiz, Elizabeth Cary, 219–20
Agassiz, Louis, 73, 220
agnotology, 231
AI (artificial intelligence), 94, 115–16
air travel, 213–14
Alfassa, Mirra, 98
alivehoods, defined, 179
American Petroleum Institute (API), 146–47
antisemitism, 44
Argentina, 97
Arnstein, Sherry: and Arnstein ladder, 206–7
Aronoff, Kate, 47
artificial intelligence (AI), 94, 115–16
attribution science, 140

Aurobindo, Sri, 98
Auroville (Tamil Nadu, India), 97–102, 99t, 101f

Bacow, Lawrence, 152, 156, 162
Baldwin, Davarian, 20, 199
Barry, John, 23, 233
Battle, Colette Pichon, 36
B Corporations, 52
Benjamin, Ruha, 230–31
bias. *See* discrimination and bias
billionaire class. *See* elites
bisexuals. *See* LGBTQ+ community
Black Americans: climate injustices and, 1; dehumanization of, 220; HBCUs and, 69; police brutality against, 45; reparations for, 74, 138; slavery and, 73–74, 116, 219–22
Black feminism, 60, 231–32
Black Lives Matter movement, 69
Bohemian Football Club (Dublin, Ireland), 239
BP partnership with Princeton University, 147
Brown University, 168
Butler, Octavia, 231–32

Campus de la Transition (France), 114–15, 115t
capitalism: academic, 20, 80, 159–68, 178, 181; climate crisis and, 78; defense of, 233; defined, 77–78; exploitative, 55;

capitalism (*cont.*)
 extractive, 10, 14, 36; in higher education, 18, 61, 78; perpetual growth model of, 178; public investments and, 110; racism and, 82; reinforcement of, 9
Catholic University of Ireland, 62
Center for Community Resilience Research Innovation and Advocacy (CCRRIA), 190
Chronicle of Higher Education, 67, 165
cisnormativity, 64
civic engagement. *See* community engagement
Civil Rights Act of 1964, 69
class. *See* social class
climate change. *See* climate crisis
climate coloniality, 82, 87–88, 216
climate crisis: capitalism and, 78; curriculum on, 103–8, 112; denial of, 37, 85, 123; as existential threat, 48; future outlook on, 236; globalization and, 200–201; greenhouse gas emissions and, 4, 15, 30, 31, 84, 213; higher education efforts to address, 82–83; intersection with other crises, 25, 36, 51, 54; perpetuation of, 87–88; resilience to, 174, 216; systemic problems related to, 55; technological fixes for, 19, 31–34, 84, 125–28, 131; vulnerabilities to, 2–5, 9, 14, 41–42, 175. *See also* fossil fuels
climate injustices: capitalism and, 78; constraints on responses to, 103; economic inequities and, 2, 31, 83; of globalization, 201; higher education and, 18, 61, 72, 82–83, 88, 205, 217; minimization of, 42, 47; paradoxes of, 95; patriarchy and, 14, 91, 133; perpetuation of, 60, 105, 131; race and, 1, 2; reparations for, 138; systemic changes for addressing, 127; worsening of, 12, 17, 23, 57, 83
climate isolationism, 32–35, 33f, 39, 83–86, 129–33, 221
climate justice: ADAPT-ing framework for, 31, 236; cooperatives and, 181; culture of care and, 228; curriculum on, 113; defined, 14; delays and resistance to, 39, 41, 90, 221; engagement with, 24, 106, 203; feminism and, 14, 32, 33, 88, 218; financial innovation and, 169–70; global solidarity and, 13–15, 21, 216; inclusiveness of, 30, 31, 36; integrated commitment to, 86, 87; knowledge creation for, 19; leveraging of disruptive events for, 189; neutrality on, 9, 47, 227; paradigm shift toward, 33f, 34, 35, 52; philanthropy and, 154; power of, 17, 34; principles of, 33f, 34, 222, 226; research and, 15, 36, 129, 132–33, 138, 146–47, 170; scale of, 12, 29; social change and, 8, 34, 47; solidarity and, 13–15, 21, 31; starting points for, 15–17; systemic change and, 31, 127, 217, 237; types of, 30; unlearning and, 97, 106, 123; well-being and, 46
climate justice universities: capitalism and, 78; as critical infrastructure, 233–37; culture of care and, 186; curriculum at, 109; distributed network approach for, 223; empowerment and, 4, 8, 94, 124, 215; funding for, 19, 82, 89; global solidarity and, 187; hope, humility, and, 8, 238; inclusivity of, 197, 232; journalistic presence of, 209; knowledge co-creation and, 57, 142, 239; objectives of, 4, 6; proposal for, 217–19; public good and, 54, 197; regenerative approach and, 205; relational knowledge and, 122; research and, 142; restructuring, 212, 216, 219; in rural communities, 198–99; as shift away from climate isolationism, 34, 221; starting points for, 15–16; as transformative framework, 90, 240; transparency and accountability in, 35, 91, 229; union, 239; unlearning and, 93, 124
Climate Justice Universities Union, 239
climate obstruction, 37–43, 81, 84, 105
Club of Rome, 112–13

collective action, 6–9, 31–36, 47, 90, 199, 203, 224–26, 236
collective good. *See* public good
collective well-being, 6, 8, 42, 71, 223
College of the Atlantic (Bar Harbor, Maine), 63
colleges. *See* climate justice universities; higher education
coloniality: climate, 82, 87–88, 216; in higher education, 18, 61, 74–77, 221, 227; indigenous cultures and, 24, 46, 75–76; othering and, 191, 196; settler colonialism, 46, 75–76; technological innovation and, 26; white supremacy and, 71. *See also* decoloniality
common good. *See* public good
community engagement: critical infrastructure for, 234; for deliberative democracy, 205–7; DRIFT and, 143–44, 210–12, 216; ecoversities and, 193–98, 216; global solidarity and, 20, 187, 202–3, 206, 213, 215; justice-centering framework for, 191–92; localism and, 195–98, 200–205, 208–9; municipalization of higher education and, 176; prioritization of, 7, 85, 226–27; public good and, 203; research and, 22, 86, 142, 144; RISE Network and, 188–90, 189f, 192, 216; in rural areas, 198–99
Connell, Raewyn, 53
Conway, Erik, 38, 79
cooperatives, 20, 180–83
corporatization of higher education, 7, 43, 78, 89, 183, 237
COVID-19 pandemic, 12, 166
critical infrastructure, 3, 5, 20, 174–76, 216, 233–37
critical university studies, 6–7, 18, 62, 63
culture: of accountability, 20, 167, 199, 229–30, 237, 238; of care, 20, 183–86, 227–28, 238; of control, 26; of economic profession, 110; of fear, 44; individualistic, 158; of institutional loyalty, 155; intercultural dialogue, 195t, 195–96; patriarchal, 49, 68, 69, 89;

of philanthropy, 159; of transparency, 167, 229; woke, 45. *See also* indigenous cultures; race and ethnicity
curriculum: banning topics from, 43, 44; on climate crisis, 103–8, 112; on climate justice, 113; of deception, 102–5, 107; decolonization of, 76; defined, 97, 102; on economics, 105–12; Eurocentric, 75; gendered nature of, 64; indigenous knowledge in, 121; outside influences on, 38, 88, 167; segregation and, 109; social change and, 104, 105; solidarity and, 119; on sustainability, 104–5, 112; unlearning and, 102–5

Daggett, Cara, 89
Daly, Hannah, 234–35
debt. *See* student debt
decoloniality: application of term, 76–77; climate justice and, 88; ecoversities and, 195t, 195–96; epistemic pluralism and, 120; indigenous languages and, 23; justice-first perspective and, 13; place-based approach and, 203; of research and curriculum, 76
dehumanization: of Black Americans, 220; coloniality and, 75, 88; disconnection and, 116, 158, 221; perpetuation of, 5, 10, 221; resistance to, 20, 202, 238
DEI. *See* diversity, equity, and inclusion
deliberative democracy, 10, 27, 205–8, 215
democracy: climate issues and, 132; deliberative, 10, 27, 205–8, 215; energy, 7, 20, 92, 159, 190; preservation of, 167; public engagement and, 106, 201; threats to, 44
DeSantis, Ron, 44–45
Design Impact Transition (DIT) platform, 212
discrimination and bias: antisemitism, 44; artificial intelligence and, 94, 116; classism, 53; in curriculum, 102; Eurocentrism, 75–76; in higher education ranking system, 179; Israeli response to

discrimination and bias (*cont.*)
Hamas attacks and, 217; minimization in Western science, 119; in research, 145–46, 150; sexism, 53, 60, 64–69, 72, 89; unlearning and, 50; wealth supremacy and, 28, 185. *See also* racism

distributive climate justice, 30

DIT (Design Impact Transition) platform, 212

diversity, equity, and inclusion (DEI), 11, 44, 69–70, 77. *See also* race and ethnicity

divestment, 88, 167, 168, 173

Doyle, Linda, 183

Dublin, Ireland, 23, 46, 154, 182–83, 237, 239

Dutch Research Institute for Transitions (DRIFT), 143–44, 210–12, 216

ecological economics, 109

ecological health, 7, 111, 128–29, 141, 151, 186, 228

ecological well-being, 123

economic inequities: calls for reversal of, 235; climate injustices and, 2, 31, 83; exnovation research and, 137; expansion and reinforcement of, 15, 41, 51; globalization and, 200, 201; higher education and, 53, 157, 164–67, 169; racial and ethnic, 71, 74; wealth caps and, 177, 182

economic infrastructure, 82

economic justice, 8, 12, 47, 52–53, 83, 169–70, 172

economics: care-based, 14, 19, 228; civic, 205–6; curriculum on, 105–12; doughnut, 110, 135–36; ecological, 107, 109; evolutionary, 111; exnovation research and, 135–38; feminist, 107, 109, 111; free market, 38, 78, 106, 201, 205; global, 48, 113, 200, 201; hegemonic power and, 48; knowledge-based, 160, 162; neoliberal, 18, 108–9, 184–85; post-growth, 14–15, 136–37; regenerative, 105–13, 136; solidarity and, 12, 14, 228; unlearning,

105–13; well-being and, 12, 19, 170; women in, 110–12. *See also* capitalism

Ecoversities Alliance, 102, 114, 193–98, 195t, 205, 216, 224–25

education. *See* higher education

elites: capture of higher education by, 6, 56, 81; financial, 78–80, 159; fundraising among, 220; philanthropy of, 156; polluter, 83, 133; solar geoengineering and, 127, 133; taxation of, 178; wealth and power concentrated among, 10–11, 41, 51. *See also* wealth

Elizabeth I (queen of England), 46, 182

emplacement, 194–97, 195t, 200

empowerment: civic renewal and, 206; climate justice universities and, 4, 8, 94, 124, 215; collective action and, 36; imagination and, 17, 230; local, 174, 187, 191–92, 200–205, 208–9, 213–16; for regenerative transformation, 210; social change and, 113

energy democracy, 7, 20, 92, 159, 190

epistemicide, 120

epistemic justice, 120–22

epistemic pluralism, 18, 120

epistemologies, 97, 119–20

Erickson, Jon, 107

ethnicity. *See* race and ethnicity

Eudora Times (newspaper), 209

Eurocentrism, 75–76

evans, tina lynn, 114

exnovation: defined, 50, 130; economic and financial, 135–38; fossil fuel and plastics, 138–42; research on, 19, 133–42, 202; shift from innovation to, 34

experiential learning, 86, 117, 194–96, 195t, 227

ExxonMobil, 163

Facer, Keri, 236

Faculty for a Future, 150, 225

fair trade learning, 117

feminism: Black, 60, 231–32; climate justice and, 14, 32, 33, 88, 218; on culture of care, 228; economics and, 107, 109,

111; justice-first perspective and, 13. *See also* gender; women
Feminist Communities for Climate Justice, 239
Fihri, Fatima al-, 45
financialization of higher education, 152–86; academic capitalism and, 20, 80, 159–68, 170, 178, 181; academic-industrial complex and, 81–82; acceleration of, 79–80, 161; culture of care and, 183–86; defense of, 233; defined, 78–79, 157, 159; disconnection and, 158; fundraising and, 154, 155, 175, 183, 220; implications of, 50–51, 63, 80, 84–85, 159, 160f, 215; philanthropy and, 81–82, 152–56, 159; public good and, 168–73; regenerative structures, 19, 158, 179; research systems and, 147; resistance to, 7, 20, 159–68; restructuring for wealth distribution, 173–80; salaries of university presidents, 165–66, 182; student debt and, 7, 153, 154, 159, 165
Florida, state university system in, 44–45
Floyd, George, 69–70
fossil fuels and fossil fuel industry: air travel and, 213; attribution science and, 140; climate obstruction by, 38–40, 81; divestment movement, 88, 167, 168, 173; emissions from, 30, 33, 39; energy democracy and, 7; extraction of, 14, 15, 38, 77; petromasculinity and, 89; phasing out, 10, 19, 37–40, 128, 133–35, 138–40, 142; power and wealth of, 81, 128, 167; reliance on, 34, 37, 39, 40, 85, 123, 167; reparations for climate damage, 138; university partnerships with, 38, 40, 43, 79, 81, 85, 88–89, 147, 163, 173
Foxx, Virginia, 44
Francis, Dania, 74
Friere, Paulo, 114
fundraising, 154, 155, 175, 183, 220

Gallagher, Patrick, 180–81
Gates, Bill, 132
gay people. *See* LGBTQ+ community
Gaza, 168, 234
GDP (gross domestic product), 109
gender: cisnormativity and, 64; educational restrictions on, 45; of leaders in higher education, 49, 64, 67–68, 68f; nonbinary individuals, 48, 49; oppression based on, 42, 60, 64; sexism, 53, 60, 64–69, 72, 89; violence and, 65–69. *See also* feminism; women
generosity, 19, 92, 152, 156–59, 162, 163, 192
genocide, 116, 234
geoengineering, 126–29, 132–33
Gibson-Graham, J. K., 212
globalization, 161, 198, 200–205
global solidarity: climate justice and, 13–15, 21, 216; community engagement and, 20, 187, 202–3, 206, 213, 215; higher education institutions and, 223–24, 226, 239
global warming. *See* climate crisis
Gluckman, Nell, 67
Graham, Amanda, 27–28
Gramsci, Antonio, 48
greenhouse gas emissions, 4, 15, 30, 31, 84, 213
greenwashing, 77, 83, 229
Griffin, Kenneth, 152, 156
gross domestic product (GDP), 109
Gutmann, Amy, 165–66

Hamas attacks on Israel (2023), 44, 168, 217
Hamer, Fannie Lou, 232
harassment. *See* sexual harassment
Harvard University: climate and sustainability work at, 219, 221; endowment for, 74, 156, 176; Legacy of Slavery Initiative, 73–74, 219–22; Palestinian supporters at, 217, 219, 222; philanthropy at, 74, 152–54, 156; sexual harassment at, 58–60, 156; solar geoengineering research program at, 132
HBCUs (historically Black colleges and universities), 69

health: ecological, 7, 111, 128–29, 141, 151, 186, 228; economic and financial, 169, 170; exnovation research and, 137; fossil fuels and, 39; global, 52, 132; inequities related to, 14, 69, 71, 83, 132; mental, 11, 82, 90, 157, 159, 172, 205; planetary, 6, 46, 108, 142, 144–45; political, 170; public, 135, 143, 148, 209; SDGs on, 54; wealth and, 179; well-being in relation to, 104

hegemony, 47–50, 53

Hernandez, Mariu, 97

Higgins, Michael D., 106

higher education: academic freedom and, 22, 46, 56, 61, 80, 222, 234–35; access to, 15, 71–72, 80, 165, 170–71; capitalism in, 18, 61, 78; climate injustices and, 18, 61, 72, 82–83, 88, 205, 217; climate obstruction in, 37–43, 84; collective reimagining of, 6–9, 17, 60–61; coloniality in, 18, 61, 74–77, 221, 227; cooperative model for, 20, 180–83; corporatization of, 7, 43, 78, 89, 183, 237; as critical infrastructure, 3, 5, 20, 174–76, 216, 233–37; critical university studies, 6–7, 18, 62, 63; disciplinary boundaries in, 4, 26, 42, 55, 103; distributed networks of, 20, 174, 176–77, 185, 197–99, 204, 223; ecoversities, 102, 114, 193–98, 195t, 205, 216, 224–25; gatekeeping in, 6, 66, 72, 147; gender-based violence in, 65–69; global, 49, 65, 203–5, 223–24; HBCUs, 69; history of, 45–46, 67; inequities and, 53, 56, 90, 157, 161, 164–67, 169, 233; Irish academic conferences, 22–25; local community relationships with, 86–87; municipalization of, 175–76, 178, 202; neutrality in, 9, 11–12, 46; paradigm shifts in, 4, 6, 8, 35, 56–57; paradox of, 27–28; patriarchy in, 18, 58–61, 64–69, 72, 84, 89, 221, 227; pedagogies in, 18, 64, 76, 97, 113–18, 115t, 219; political power in, 43–47; power of, 4, 8, 17, 25, 43, 56, 209, 217, 234, 238; public funding for, 79, 81, 161–62, 165, 173–76, 178; as public good, 20, 52–54, 87, 90–91, 95, 157, 159, 225–26, 234; purpose and values of, 20, 226–28; racism in, 18, 53, 61, 67, 69–74, 221, 237; ranking systems in, 36, 163, 178–79; sexism in, 53, 60, 64–69, 72, 89; societal role of, 35, 40, 53–55, 62, 93, 234; tax-exempt status of institutions, 87, 172, 215; transformative lens for, 25–29, 36; as underleveraged resource, 3, 53, 56, 192, 233; well-being in, 89–90, 93, 124; for women, 183, 220. *See also* climate justice universities; community engagement; curriculum; financialization of higher education; knowledge; leaders and leadership; research; unlearning; *and specific institutions*

himpathy, defined, 66

historically Black colleges and universities (HBCUs), 69

hooks, bell, 114, 232

hope, 8, 57, 124, 138, 213, 232, 233, 238

humility, 8, 93, 128, 133, 144, 194, 238

Hurricane Katrina (2005), 1–3, 15

Hurricane Maria (2017), 118, 187–88, 216

IAU (International Association of Universities), 224

identity taxation, 70

ignorance, 230–31

imagination: Black feminist, 231–32; collective, 9, 16, 42, 93; constraints on, 84, 85, 208, 232; Ecoversities Alliance and, 102, 225; empowerment and, 17, 230; expansion of, 7, 128, 205; for justice, 230–33; power of, 8, 230, 232

indigenous cultures: coloniality and, 24, 46, 75–76; knowledge of, 75, 120–22, 145, 179, 194; languages of, 24; reciprocity in, 19, 118, 157; reparations for, 138

individualism: arrogance and, 128; competition and, 55, 62, 126, 158, 184, 203; knowledge hierarchies and, 62, 122; reinforcement of, 5, 10; structural racism and, 70; technocratic, 84, 217; unlearning, 118 inequities: climate isolationism and, 86, 131; fair trade learning and, 117; health-related, 14, 69, 71, 83, 132; higher education and, 53, 56, 90, 157, 161, 164–67, 169, 233; perpetuation of, 123, 237; political, 31, 41; racial, 2, 69, 71, 74, 77; reduction of, 34, 170, 205; SDGs on, 54; structural forces and, 185. *See also* economic inequities

infrastructure: critical, 3, 5, 20, 174–76, 216, 233–37; economic, 82; energy, 134; funding for, 87; physical, 80, 82, 160; political, 206; social, 20, 85, 143, 199, 206, 214–15, 235

innovation: in Auroville, 98; cooperatives and, 181; defined, 50, 130, 134; financial, 138, 169–71; pedagogical, 18, 114; in private sector, 110; shift to exnovation from, 34; social, 26, 27, 33, 82, 84–86, 134; technological, 26–27, 33–34, 82, 84–86, 127–28, 131, 133–34; in university-community relationships, 190, 210

intercultural dialogue, 195t, 195–96
intergenerational justice, 30
Intergovernmental Panel on Climate Change (IPCC), 95
International Association of Universities (IAU), 224
Israel: Hamas attacks on (2023), 44, 217; settler colonialism in, 75; university activism during war in Gaza, 168, 234

Jain, Manish, 100, 102, 193
journalism, 66, 88–89, 208–9
justice-centering relationship framework, 191–92
justice-first perspective, 9–13

Kaba, Mariame, 238
Kelly, Marjorie, 28, 185
Kelton, Stephanie, 110–11
Kendi, Ibram X., 47
Keynes, John Maynard, 110
Kimmerer, Robin Wall, 118, 157, 180

Kirby, Peadar, 24
Klein, Naomi, 200
knowledge: co-creation of, 19, 57, 130–31, 142–45, 175, 210–11, 219, 239; commodification of, 80, 159, 173; contextualization of, 223; democratization of, 94; dissemination of, 9, 18, 43, 55–56, 93–97, 218, 235; dominant frameworks of, 95; economies of, 160, 162; epistemologies and, 97, 119–20; Eurocentric, 75–76; geopolitics of, 120; hierarchies of, 35, 62, 102, 122, 193, 197; imbalance in university-community interactions, 191; indigenous, 75, 120–22, 145, 179, 194; local, 75, 100, 117, 120, 145, 179, 198, 209; misinformation, 10, 143, 168, 208; neutrality of, 46, 150; organization of, 26, 93; power of, 41, 46, 93, 128–29, 230–31; redistribution of, 6, 91, 158, 202; relational, 19, 103, 116–18, 122, 177, 203, 227; wealth, power, and, 6, 9, 17–18, 40–41, 41f, 43, 46–47, 61, 97, 120, 185, 209. *See also* higher education; research; unlearning

knowledge to action gap, 130, 142
Koch family, 85, 147, 167–68
Kyoto Protocol (1997), 146

Lachapelle, Paul, 229
Lasagna, Louis, 149
leaders and leadership: authoritarian, 45, 55; for climate justice, 12; curricular integration and, 105; demographics of, 49, 64, 67–68, 68f; financialization and, 50, 162; hegemonic power and, 48; learning-centered, 229; legitimacy of, 222; patriarchy and, 49, 84; political intimidation of, 43–44; for racial justice, 74; salaries of, 165–66, 182
lesbians. *See* LGBTQ+ community
LGBTQ+ community, 45, 64, 66, 121
localism, 195–98, 200–205, 208–9, 212–16
Lorde, Audre, 60
loyalty, 152, 155–59, 162, 192
Lutunatabua, Thelma Young, 20–21

Machado de Oliveira, Vanessa, 121–22
Mandela, Nelson, 233
Mann, Kate, 66
market fundamentalism, 38, 106
Mazzucato, Mariana, 110
McGeown, Calum, 22
men. *See* gender; patriarchy
mental health, 11, 82, 90, 157, 159, 172, 205
#MeToo movement, 67
Michaels, David, 40
Mignolo, Walter, 120
militarization, 33, 55
minorities. *See* race and ethnicity
misogyny. *See* sexism
Mondragon University (Spain), 181–82, 237
Monticelli, Lara, 111
Moore, Thom, 72
municipalization of higher education, 175–76, 178, 202
mutual aid, 117–18, 184

National Labor Relations Act, 166
neoliberalism, 18, 54, 88, 108–9, 184–85, 202
Netherlands, 179
Newman, John Henry, 62
new municipalism, 175–76, 202
news deserts, 208, 209
New York Times, 163, 178, 201
Northeastern University (Boston): Board of Trustees, 163; community-engaged research at, 144; DEI initiatives at, 70; economic inequities within, 165–67; fossil fuel phaseout research at, 139–40; Giving Day at, 152–53, 183–84; as global university system, 203–4; School of Public Policy and Urban Affairs, 10; *The Scope* digital magazine and, 209; tuition fees at, 165

Ó hÓgartaigh, Ciarán, 24
oppression: accountability for, 149; climate justice and, 29, 34; dismantling systems of, 11, 121; exposure through research, 149; gender-based, 42, 60, 64; interlocking systems of, 82, 83; of Palestinians, 222; pedagogies of, 113–16; reinforcement of, 77, 91; solidarity against, 118; structural, 232; unlearning, 195
Oreskes, Naomi, 38, 79
Ortiz-Garcia, Cecilio, 188–90
othering, 158, 191, 196, 202

Palestinian people, 44, 217, 219, 222
patriarchy: artificial intelligence and, 94; capitalism and, 82; climate injustices and, 14, 91, 133; hegemony of, 48, 49; in higher education, 18, 58–61, 64–69, 72, 84, 89, 221, 227; himpathy and, 66; petromasculinity and, 89; research and, 32, 49, 133; technological innovation and, 26
pedagogies, 18, 64, 76, 97, 113–18, 115t, 219
Perez, Carlota, 111
Perez-Lugo, Marla, 188–90
petromasculinity, 89
philanthropy, 74, 81–82, 127, 132, 148, 152–56, 159
physical infrastructure, 80, 82, 160
PILOT (payment in lieu of taxes) programs, 87
planetary health, 6, 46, 108, 142, 144–45
planetary limits, 19, 107–8, 110, 112, 134–36, 184
Planetary Limits Academic Network (PLAN), 225
A Plastic Planet, 141–42
plastics exnovation, 19, 138–39, 141–42
political health, 170
political infrastructure, 206
polluter elites, 83, 133
polycrisis, 11, 25, 29, 36, 238
polygenesis, 220
Positive Money, 170
poverty, 54–55, 107, 185, 191
Powell, Jerome, 108
power: abuses of, 65, 67, 91; access to, 150; accumulation of, 79, 129; ADAPT-ing framework and, 31, 236; climate iso-

lationism and, 131; of climate justice, 17, 34; collective, 158, 224; colonial, 75, 87–88; concentration of, 5–6, 9–11, 14, 18, 29, 34, 41, 51, 55, 88, 132, 148, 159, 171; as constraint on academic inquiry, 218; of cooperatives, 180; diversifying, 12–13; dynamics of, 32, 48, 63, 75, 120, 152–53, 169, 211; economic, 7, 29; of elites, 133; exploitative, 11, 13, 75; of fossil fuel industry, 81, 128, 167; hegemonic, 47–50; of higher education, 4, 8, 17, 25, 43, 56, 209, 217, 234, 238; of imagination, 8, 230, 232; imbalances of, 53, 87, 117; institutional, 24; of knowledge, 41, 46, 93, 128–29, 230–31; of legacy, 219–22; leveraging, 25, 29, 43, 45–47, 56, 125, 153, 167, 173, 223, 235; political, 7, 29, 43–47; of private sector, 38; redistribution of, 6, 7, 91, 132, 158, 170, 201, 207; regenerative, 20, 238; structures of, 11, 13, 32, 60, 75, 77, 95, 126, 129, 144, 191, 221, 228–29; in university-community relationships, 196; of unlearning, 92, 123, 124; wealth, knowledge, and, 6, 9, 17–18, 40–41, 41f, 43, 46–47, 61, 97, 120, 185, 209. *See also* empowerment

prejudice. *See* discrimination and bias

Princeton University, 79, 147, 155

procedural climate justice, 30

public good: climate justice universities and, 54, 197; collective action for, 6, 36, 47; community engagement and, 203; erosion of sense of, 221; higher education as, 20, 52–54, 87, 90–91, 95, 157, 159, 225–26, 234; justice-first perspective and, 10–12; prioritization of, 5, 7, 85, 196; private interests as threat to, 167–68; reclaiming financial flows for, 168–73; research for, 51, 145, 147–51; social innovation and, 82, 134; systemic change for, 10; tax contributions for, 178

public health, 135, 143, 148, 209

purpose-driven organizations, 227

queer people. *See* LGBTQ+ community

Rabe, Chris, 113

race and ethnicity: climate injustices and, 1, 2; DEI initiatives, 11, 44, 69–70, 77; educational restrictions on, 45; exclusion from university knowledge systems, 120–21; inequities based on, 2, 69, 71, 74, 77; of leaders in higher education, 67, 68f; polygenesis and, 220; violence and, 45, 69–70; wealth gap and, 74. *See also* Black Americans; indigenous cultures

racial justice, 47, 69, 74, 77, 154, 218

racism: ADAPT-ing framework and, 31, 236; artificial intelligence and, 94; capitalism and, 82; environmental, 71; in higher education, 18, 53, 61, 67, 69–74, 221, 237; neutrality on, 47; segregation, 69, 73, 220; structural, 42, 69–74; white supremacy, 28, 70–72, 82, 89, 91, 185, 220

rape. *See* sexual violence

Raworth, Kate, 110, 135–36

reciprocity, 19, 104, 116–18, 157

recognition justice, 30

REF (Research Excellence Framework), 161

Renner, Edward, 72

reparative justice, 74, 120–22

research, 125–51; agendas for, 28, 79, 126–28, 133–40, 148–51, 163, 167; bias in, 145–46, 150; climate justice and, 15, 36, 129, 132–33, 138, 146–47, 170; climate obstruction and, 37, 38; collaborative, 10, 19, 81, 130, 139–40, 142–44, 149; community-engaged, 22, 86, 142, 144; constraints on, 42, 63, 71, 89, 125; corporate influences on, 145–48, 163, 167; damage-centered, 149; decoloniality and, 76; on exnovation, 19, 133–42, 202; funding for, 40, 74, 80–82, 85–89, 132, 146–50, 161, 164, 220; global solidarity and, 15; patriarchy and, 32, 49, 133; for public good, 51, 145, 147–51;

research (cont.)
regenerative, 142, 144–45, 151; societal benefits of, 27; on solar geoengineering, 126–28, 132–33; sustainability in, 105
Research Excellence Framework (REF), 161
research-for-profit model, 145–48
"resist, reclaim, and restructure" framework, 19–20, 88, 159
RISE Network, 188–90, 189f, 192, 216
Roberts-Gregory, Frances, 35
Robinson, Joan, 111–12
Robinson, Mary, 32

Scholars for a New Deal for Higher Education (SFNDHE), 225
Scientist Rebellion, 225
Scientists for Global Responsibility, 225
The Scope (digital magazine), 209
SDGs (Sustainable Development Goals), 54–55, 104–5
SDSN (Sustainable Development Solutions Network), 112
segregation: curricular, 109; racial, 69, 73, 220; of society, 51
settler colonialism, 46, 75–76
sexism, 53, 60, 64–69, 72, 89
sexual harassment, 58–60, 64–66, 68
sexual orientation. *See* LGBTQ+ community
sexual violence, 64, 66, 69, 91
SFNDHE (Scholars for a New Deal for Higher Education), 225
Sheehan, Sinéad, 22
Shiva, Vandana, 198
slavery, 73–74, 116, 219–22
Smith, David, 202
social change: climate justice and, 8, 34, 47; climate obstruction and, 40; collective imaginations about, 9; curriculum and, 104, 105; engagement in, 113; higher education and, 53; imagination and, 230; investments in, 86; justice-first perspective and, 10; research and, 151; systemic, 25–27; Think-tank for Action on Social Change, 106; transformative, 4–6, 8, 25–26, 33, 51, 55, 84, 132, 239
social class: discrimination based on, 53; hegemonic power and, 48. *See also* elites
social footprint mapping, 20, 199–200
social infrastructure, 20, 85, 143, 199, 206, 214–15, 235
social justice: advocacy for, 8, 61, 228; climate issues and, 10, 83, 132; constraints on, 61, 184; fair trade learning and, 117; investments in, 85; neutrality on, 47; public good and, 52, 53; research agendas and, 135; solidarity and, 119; universities and, 176, 218
solar geoengineering, 126–29, 132–33
solidarity: climate justice and, 13–15, 21, 31; cooperatives and, 181; economics and, 12, 14, 228; inquiry in, 194–96, 195t; with Palestinians, 217; teaching, 118–19, 124. *See also* global solidarity
Solnit, Rebecca, 20–21
Starr, G. Gabrielle, 94
St. Clair, Ralf, 226
Stein, Sharon, 236
student debt, 7, 153, 154, 159, 165
sustainability: commitments to, 4, 56, 227; curriculum on, 104–5, 112; DRIFT and, 143, 210–12; ecoversities and, 194; globalized, 55; greenwashing and, 229; Harvard's work on, 219, 221; pedagogies of, 114; strategic initiatives for, 23; transitions to, 28, 92, 143, 210–12; unlearning and, 92
Sustainable Development Goals (SDGs), 54–55, 104–5
Sustainable Development Solutions Network (SDSN), 112
Synergy Grant program (European Research Council), 137
systemic change: in academic culture, 23; challenges of, 16, 95, 132; climate justice and, 31, 127, 217, 237; dimensions of,

228–29; for public good, 10; social change as, 25–27

Tamale, Sylvia, 150
TASC (Think-tank for Action on Social Change), 106
taxation: exemptions for higher education institutions, 87, 172, 215; identity taxation, 70; for public good, 177; of wealth, 170, 178
technology: artificial intelligence, 94, 115–16; for carbon capture and removal, 33, 39; climate crisis fixes and, 19, 31–34, 84, 125–28, 131; innovation in, 26–27, 33–34, 82, 84–86, 127–28, 131, 133–34; solar geoengineering, 126–29, 132–33
"Think Globally, Act Locally" concept, 212–16
Think-tank for Action on Social Change (TASC), 106
third-level education. *See* higher education
Toffler, Alvin, 128–29
tokenism, defined, 207
Tooze, Adam, 25
transgender people. *See* LGBTQ+ community
transparency: climate justice and, 35; culture of, 167, 229; financial, 222, 229; importance of, 91; power and, 56; resistance to, 237; self-reflection and, 12, 72–73, 235
Trinity College (Hartford, CT), 199
Trinity College Dublin, 23, 46, 182–83, 237
Tuck, Eve, 76, 149

United Nations (UN): climate conferences, 32, 39, 139; Framework Convention on Climate Change, 15; Intergovernmental Panel on Climate Change, 95; Sustainable Development Goals, 54–55, 104–5; Sustainable Development Solutions Network, 112
United Nations Education, Scientific and Cultural Organization (UNESCO), 55–56, 224
universities. *See* climate justice universities; higher education
University of al-Quaraouiyine (Morocco), 45
University of Barcelona, 112
University of Bologna (Italy), 45
University of Puerto Rico, 118, 187–88
University of Texas at Rio Grande Valley (UTRGV), 188–90
University of the People (UoPeople), 170–71
UnKoch My Campus, 167–68
unlearning, 92–124; artificial intelligence and, 94; in Auroville, 97–102, 99t, 101f; climate justice and, 97, 106, 123; curriculum of deception and, 102–5; defined, 50, 92; economics and, 105–13; in ecoversities, 195; epistemological hierarchies and, 119–20; individualism, 118; as liberation and justice, 122–24; as mechanism for change, 95; pedagogies of oppression and, 113–16; power of, 92, 123, 124; resistance to, 96; sustainability and, 92; value of, 18–19, 92–93
UoPeople (University of the People), 170
Utrecht University, 179
UTRGV (University of Texas at Rio Grande Valley), 188–90

van't Land, Hilligje, 224
violence: exploitative, 59; gender-based, 65–69; genocide, 116, 234; geopolitical, 54; race-based, 45, 69–70; sexual, 64, 66, 69, 91; state, 222

Waddock, Sandra, 228
wealth: accumulation of, 9, 34, 41, 46, 79, 129, 154, 157, 162, 171; of alumni, 152; caps on, 137, 171, 176, 178, 182; climate isolationism and, 131; collective, 158; community, 20, 27, 174, 176; concentration of, 5–6, 9–11, 14, 18, 29, 34, 41,

wealth (*cont.*)
51, 55, 88, 148, 159, 171; conceptual restructuring, 179; as constraint on academic inquiry, 218; of fossil fuel industry, 128, 167; inequities related to, 69, 191; knowledge, power, and, 6, 9, 17–18, 40–41, 41f, 43, 46–47, 61, 97, 120, 185, 209; racial wealth gap, 74; redistribution of, 6, 19–20, 35, 91, 132, 158, 169–70, 172–80, 201; from slave trading, 73; taxation of, 170, 172, 178; of universities and university leaders, 166, 221–22. *See also* elites

wealth supremacy, 28, 185

well-being: assessment of, 108; climate justice and, 46; collective, 6, 8, 42, 71, 223; coloniality and, 88; cooperatives and, 181; ecological, 123; economic, 12, 19, 170; exnovation research and, 137; health in relation to, 104; in higher education, 89–90, 93, 124; of local communities, 196; prioritization of, 7, 34, 111, 172, 174, 176; research and, 143; undermining, 89–91

Western science, 119–20

White-Newsome, Jalonne, 31, 236, 237

white supremacy, 28, 70–72, 82, 89, 91, 185, 220

woke culture, defined, 45

women: in economics profession, 110–12; exclusion from university knowledge systems, 121; higher education for, 183, 220; as leaders in higher education, 68, 68f; sexual harassment of, 58–60, 64–66, 68; sexual violence against, 64, 66, 69, 91; university programs in support of, 49. *See also* feminism; gender

worker-owned organizations, 20, 180–83

Yale University, 163
Yang, K. Wayne, 76, 149

Zinn, Howard, 12

More on Higher Ed from HOPKINS PRESS

UNIVERSITIES ON FIRE

Higher Education in the Climate Crisis

BRYAN ALEXANDER

"We all owe an enormous debt of gratitude to Bryan Alexander for his efforts to provide a common framework and scenarios to enable universities leaders to develop proactive strategies and policies to meet current and future climate change challenges."

—*Inside Higher Ed*
Joshua Kim, coauthor of *Learning Innovation and the Future of Higher Education*

 | PRESS.JHU.EDU |

Milton Keynes UK
Ingram Content Group UK Ltd.
UKHW031318071224